Springer Theses

Recognizing Outstanding Ph.D. Research

For further volumes:
http://www.springer.com/series/8790

Aims and Scope

The series "Springer Theses" brings together a selection of the very best Ph.D. theses from around the world and across the physical sciences. Nominated and endorsed by two recognized specialists, each published volume has been selected for its scientific excellence and the high impact of its contents for the pertinent field of research. For greater accessibility to non-specialists, the published versions include an extended introduction, as well as a foreword by the student's supervisor explaining the special relevance of the work for the field. As a whole, the series will provide a valuable resource both for newcomers to the research fields described, and for other scientists seeking detailed background information on special questions. Finally, it provides an accredited documentation of the valuable contributions made by today's younger generation of scientists.

Theses are accepted into the series by invited nomination only and must fulfill all of the following criteria

- They must be written in good English.
- The topic should fall within the confines of Chemistry, Physics, Earth Sciences, Engineering and related interdisciplinary fields such as Materials, Nanoscience, Chemical Engineering, Complex Systems and Biophysics.
- The work reported in the thesis must represent a significant scientific advance.
- If the thesis includes previously published material, permission to reproduce this must be gained from the respective copyright holder.
- They must have been examined and passed during the 12 months prior to nomination.
- Each thesis should include a foreword by the supervisor outlining the significance of its content.
- The theses should have a clearly defined structure including an introduction accessible to scientists not expert in that particular field.

Jaroslav Haas

Symmetries and Dynamics of Star Clusters

Doctoral Thesis accepted by
the Charles University in Prague, Czech Republic

Author
Dr. Jaroslav Haas
Faculty of Mathematics and Physics
Astronomical Institute
Charles University in Prague
Prague
Czech Republic

Supervisor
Dr. Ladislav Šubr
Faculty of Mathematics and Physics
Astronomical Institute
Charles University in Prague
Prague
Czech Republic

ISSN 2190-5053 ISSN 2190-5061 (electronic)
ISBN 978-3-319-37857-2 ISBN 978-3-319-03650-2 (eBook)
DOI 10.1007/978-3-319-03650-2
Springer Cham Heidelberg New York Dordrecht London

© Springer International Publishing Switzerland 2014
Softcover reprint of the hardcover 1st edition 2014
This work is subject to copyright. All rights are reserved by the Publisher, whether the whole or part of the material is concerned, specifically the rights of translation, reprinting, reuse of illustrations, recitation, broadcasting, reproduction on microfilms or in any other physical way, and transmission or information storage and retrieval, electronic adaptation, computer software, or by similar or dissimilar methodology now known or hereafter developed. Exempted from this legal reservation are brief excerpts in connection with reviews or scholarly analysis or material supplied specifically for the purpose of being entered and executed on a computer system, for exclusive use by the purchaser of the work. Duplication of this publication or parts thereof is permitted only under the provisions of the Copyright Law of the Publisher's location, in its current version, and permission for use must always be obtained from Springer. Permissions for use may be obtained through RightsLink at the Copyright Clearance Center. Violations are liable to prosecution under the respective Copyright Law.
The use of general descriptive names, registered names, trademarks, service marks, etc. in this publication does not imply, even in the absence of a specific statement, that such names are exempt from the relevant protective laws and regulations and therefore free for general use.
While the advice and information in this book are believed to be true and accurate at the date of publication, neither the authors nor the editors nor the publisher can accept any legal responsibility for any errors or omissions that may be made. The publisher makes no warranty, express or implied, with respect to the material contained herein.

Printed on acid-free paper

Springer is part of Springer Science+Business Media (www.springer.com)

Supervisor's Foreword

In an attempt to understand nature that surrounds us, theory, that encodes the laws of nature into mathematical formulae and experiment that discovers behaviour of the real world, need to go hand in hand. We find ourselves in a somewhat specific situation when dealing with processes that occur on scales exceeding our laboratories. This is the case of astronomy, which provides us usually with just short snapshots of systems vastly exceeding our local means of measure of space and time. The role of the theoretical part—the astrophysics—is then quite often to make a movie from a single snapshot, i.e. to deduce evolution of the observed systems on timescales of millions of years.

Galaxies are huge structures, several tens of thousands of parsecs across, formed by gas and hundreds of billions of stars. Only the innermost regions of radii up to few parsecs are usually called the galactic nuclei. Considering their spatial extent and enclosed mass, they represent just a tiny seed in comparison to their host systems. However, they are remarkable due to extreme physical conditions in them—they host central supermassive black holes of mass ranging from millions to billions of Solar masses which are surrounded by star clusters with densities exceeding million times the stellar density in the Solar neighbourhood. The unique conditions of galactic nuclei definitely deserve attention by themselves. Nevertheless, they also play an important role in a wider context. It is known that galactic nuclei keep imprints of the galactic evolution and, therefore, they are a kind of tracers of the evolution of the Universe. They are not only passive tracers; from time to time, when a sufficient amount of gas is pushed towards the supermassive black hole, an episode of so-called active galactic nucleus is triggered. The hot gas falling onto the black hole overshines the host galaxy by no less than an order of magnitude and this radiation strongly affects even the outer parts of the galaxy.

The centre of our Galaxy—The Milky Way—is the nearest galactic nucleus which can be observed with by far the best resolution. In spite of that it does not belong to the category of active galactic nuclei, it hosts all components typical for them: a supermassive black hole, very dense star cluster and a considerable amount of gas. Recent observations that allowed us to follow motions of stars and gas in the vicinity of the black hole in 'real time' revealed several mysterious facts about the Galactic Centre. In particular, discovery of several hundreds of very young stars which orbit around the supermassive black hole in distances between a few milliparsecs to about one-half of parsec was a big surprise and their origin is still unclear.

The Ph.D. thesis of Jaroslav Haas addresses the mystery of young stars in the Galactic Centre from a theoretical point of view. He stresses that their current kinematic state is definitely different from the initial one and, therefore, it can be used for determination of their origin only with caution. The non-trivial link between initial and current state of the young stars lies in their dynamical and internal evolution from the time of their birth, several millions years ago, till now. Numerical modelling of the dynamics of young stars in the Galactic Centre which is driven by mutual gravitational interaction of various constituents (supermassive black hole, dense star cluster and massive molecular torus) is the key subject of Mr. Haas' thesis. He extends and further develops older models which assume that the young stars have been born in a single gaseous disc. Then, their orbital evolution has been considerably affected by gravity of a massive molecular torus, which encircles the region under consideration. Both direct *N*-body simulations and semi-analytic approach are used to show that such a setup may lead to the state observed nowadays. The original disc tends to be partially destroyed by the gravity of the molecular torus, while mutual gravitational interaction of individual young stars keeps the 'core' of the disc together. It also shows that the residual disc tends to change its orientation towards perpendicular with respect to the molecular torus. Such an orientation is consistent with current observations and, therefore, it may be considered an indirect proof of quality of the model.

Yet another effect, which has been discussed in the thesis, is mutual gravitational interaction of young stars and a much numerous population of late-type stars. In spite of that the role of this interaction is not fully understood yet, the general message from the thesis is correct at high confidence level: the disc-like stellar structure embedded in a spherical stellar cluster leads to formation of global lopsided mode—a kind of perturbation to the mean potential—which is capable to affect individual stellar orbits in a substantial way. The most apparent consequence is secular growth of their eccentricities which may lead to tidal disruptions of the stars by the supermassive black hole.

The Ph.D. thesis of Jaroslav Haas represents an important step forward in our understanding of evolution of the Galactic nucleus, shedding new light on the role of gravitational interactions among its different constituents.

Prague, 25th June 2013 Ladislav Šubr

Preface

Over the past decades, both theoretical and observational efforts have led to a common view that most galactic nuclei host a massive central body, presumably a supermassive black hole (hereafter SMBH). It has been further accepted that galactic nuclei contain extremely dense star clusters that belong to the very old and only slowly evolving stellar population from the more distant parts of galaxies. In the first approximation, these star clusters may be considered roughly spherically symmetric. Their dynamical evolution in the potential of the central SMBH has been studied by various authors in the past, starting with the series of papers by Peebles (1972a, b) and Bahcall and Wolf (1976, 1977).

One specific target of such investigations has always been, due to its proximity, the centre of our own Galaxy. Surprisingly, the observations of this region that have been carried out during the last 20 years have established that, in addition to the old spherical cluster, the closest vicinity of the central SMBH is also occupied by very young stars (Allen et al. 1990; Genzel et al. 2003; Ghez et al. 2003, 2005; Paumard et al. 2006; Bartko et al. 2009, 2010). Moreover, it has also turned out that the spatial configuration of a substantial subset of these young stars is rather unexpected, in particular, many of them appear to belong to a coherently rotating disc-like structure identified for the first time by Levin and Beloborodov (2003). The dynamical evolution of such a stellar disc has been analysed by Alexander et al. (2007) and Cuadra et al. (2008) by means of numerical modelling. These works, however, are based on several simplifications. Among others, the most limiting one seems to be the approximation of an isolated stellar system, i.e. a system that is not influenced by any other sources of gravity except for its own and the central SMBH. Other works include the perturbative influence of a possible second stellar disc (Nayakshin et al. 2006; Löckmann et al. 2008; Löckmann and Baumgardt 2009; Löckmann et al. 2009) or the old spherical star cluster (Kocsis and Tremaine 2011).

A stellar disc similar to the one detected in the Galactic Centre has also been reported in the central parts of our neighbouring galaxy M31 (Bender et al. 2005; Lauer et al. 2012). Hence, it appears that such structures might represent generic component of galactic nuclei. In this thesis, we thus attempt to broaden the previous analyses and further investigate the evolution of the initially thin stellar discs around the SMBH. By means of numerical N-body modelling, we include the perturbative influence of an extended spherically symmetric star cluster (Chap. 2).

We discuss the case when the cluster is emulated by a predefined analytic potential in contrast to the case when it is treated in the full N-body way, as a large number of gravitating stars. In the later one, our results reveal a significant impact of the cluster gravity upon the evolution of the embedded disc. Furthermore, we consider the perturbative gravitational influence of a distant axisymmetric source which approximates massive gaseous torus, another widely expected component of the active galactic nuclei. We develop a simple semi-analytic model for this setting (Chap. 3) and apply this model to the young stellar system in the Galactic Centre, confronting the results with direct numerical N-body calculations (Chap. 4).

Before we start with the scientific part of this thesis, there are a few people I would like to thank. First of all, I thank my supervisor Ladislav Šubr for having patience with me, which might have not been always easy. I thank Marek Wolf, the director of the Astronomical Institute of the Charles University, for providing me a chair, a table and other necessities; David Vokrouhlický for his invaluable help with the research presented in this thesis; Hana Mifková, the Institute's secretary, for feeding me with a tremendous amount of homemade cakes; Attila Mészáros for all the 'PŠMs', political discussions and jokes; Petr Pokorný and Josef Hanuš for unplugging and hiding my laptop battery, dismantling my chair, continuously displacing or hiding all of my stuff and other pleasantries and Michal Švanda for scaring me in the Institute's corridors. I would like to thank also Pavel Kroupa from the Argelander Institute for Astronomy of the University of Bonn for being a kind host during my stay there and giving me much useful scientific advice and Sverre J. Aarseth from the Institute of Astronomy of the University of Cambridge for all the delicious Scottish cookies and the optimized version of his NBODY6 code without which this thesis would never exist. But most of all, I wish to thank my future wife, Jana, for being my future wife.

In this book version of my thesis, I have corrected several typos and misleading expressions and sentences.

References

Alexander R. D., Begelman M. C., Armitage P. J., 2007, ApJ, 654, 907
Allen D. A., Hyland A. R., Hillier D. J., 1990, MNRAS, 244, 706
Bahcall J. N., Wolf R. A., 1976, ApJ, 209, 214
Bahcall J. N., Wolf R. A., 1977, ApJ, 216, 883
Bartko H. et al., 2009, ApJ, 697, 1741
Bartko H. et al., 2010, ApJ, 708, 834
Bender R. et al., 2005, ApJ, 631, 280
Cuadra J., Armitage P. J., Alexander R. D., 2008, MNRAS, 388, L64
Genzel R. et al., 2003, ApJ, 594, 812
Ghez A. M. et al., 2003, ApJ, 586, L127
Ghez A. M., Salim S., Hornstein S. D., Tanner A., Lu J. R., Morris M., Becklin E. E., Duchêne G., 2005, ApJ, 620, 744
Kocsis B., Tremaine S., 2011, MNRAS, 412, 187
Lauer T. R., Bender R., Kormendy J., Rosenfield P., Green R. F., 2012, ApJ, 745, 121

Levin Y., Beloborodov A. M., 2003, ApJ, 590, L33
Löckmann U., Baumgardt H., Kroupa P., 2008, ApJ, 683, L151
Löckmann U., Baumgardt H., 2009, MNRAS, 394, 1841
Löckmann U., Baumgardt H., Kroupa P., 2009, MNRAS, 398, 429
Nayakshin S., Dehnen W., Cuadra J., Genzel R., 2006, MNRAS, 366, 1410
Paumard T. et al., 2006, ApJ, 643, 1011
Peebles P. J. E., 1972a, Gen. Rel. Grav., 3, 63
Peebles P. J. E., 1972b, ApJ, 178, 371

Abstract

We investigate the orbital evolution of an initially thin stellar disc around a supermassive black hole, considering various perturbative sources of gravity. By means of direct numerical *N*-body modelling, we first focus on the case when the disc is embedded in an extended spherically symmetric star cluster. We find that the gravitational influence of the disc triggers formation of macroscopic non-spherical substructure in the cluster which, subsequently, significantly affects the evolution of the disc itself. In another approximation, when the cluster is emulated by an analytic spherically symmetric potential, we further consider perturbative gravitational influence of a distant axisymmetric source. Using standard perturbation methods, we derive a simple semi-analytic model for such a configuration. It turns out that the additional axisymmetric potential leads to mutual gravitational coupling of the individual orbits from the disc. Consequently, the dense parts of the disc can, for some period of time, evolve coherently. Finally, we apply some of our results to the young stellar disc which is observed in the innermost parsec of the Galactic Centre.

Keywords Methods: numerical · Methods: analytical · Stellar dynamics · Celestial mechanics · Galaxy: nucleus

Acknowledgments

The author gratefully appreciates a scholarship from the Deutscher Akademischer Austausch Dienst (DAAD). This work has been partially supported by grant GACR-205/09/H033 of the Czech Science Foundation, project SVV 261301 of the Charles University in Prague, project 367611 of the Grant Agency of the Charles University in Prague and also by project LD12065 of the research programme COST CZ of the Czech Ministry of Education, Youth and Sports.

Contents

Chapter 1
Prerequisites

Before dealing with our own research, let us dedicate the first chapter of the thesis to a brief overview of dynamics of (self-) gravitating stellar systems. We start from the hierarchical three-body problem, leading then to description of relaxational processes in N-body systems. We restrain ourselves to purely a Newtonian treatment of gravity. Validity of this approach can be easily justified by comparing the strength of the most prominent first post-Newtonian relativistic correction, i.e. the pericentre advance, to the analogous process induced by another considered sources of gravity (extended spherically symmetric star cluster or external axisymmetric perturbation to the gravitational field). We further consider the stars to be point-like sources of the gravitational potential and omit any other physically relevant interactions or phenomena, such as the stellar evolution or the hydrodynamical interactions with the gaseous structures in the system.

The total gravitational potential, $\Phi_{\rm tot}(\boldsymbol{r}, t)$, in an N-body stellar system (including possible predefined potentials) can be split into two qualitatively different parts: (i) the mean, i.e., time-averaged, potential $\Phi_{\rm mean}(\boldsymbol{r})$ and (ii) the varying time-dependent component $\Phi_{\rm var}(\boldsymbol{r}, t) \equiv \Phi_{\rm tot}(\boldsymbol{r}, t) - \Phi_{\rm mean}(\boldsymbol{r})$. The particular form of these two components significantly affects the dynamical evolution of the system. In the following sections, we briefly describe the dynamics in some of the basic configurations in which $\Phi_{\rm var}(\boldsymbol{r}, t) \equiv 0$ in order to setup the basis for the research presented in the subsequent chapters.

1.1 Keplerian Elements

Let us first consider a star of mass m that is orbiting a supermassive black hole (hereafter SMBH) of mass, $M_\bullet \gg m$. In such a case, the SMBH may be assumed fixed in the centre of the system, generating the Keplerian potential $\phi_\bullet = -GM_\bullet/r$, where r denotes the distance of the star from the SMBH and G stands for the gravitational constant. The orbit of the star is a Keplerian ellipse which can be described by means

J. Haas, *Symmetries and Dynamics of Star Clusters*,
Springer Theses, DOI: 10.1007/978-3-319-03650-2_1,
© Springer International Publishing Switzerland 2014

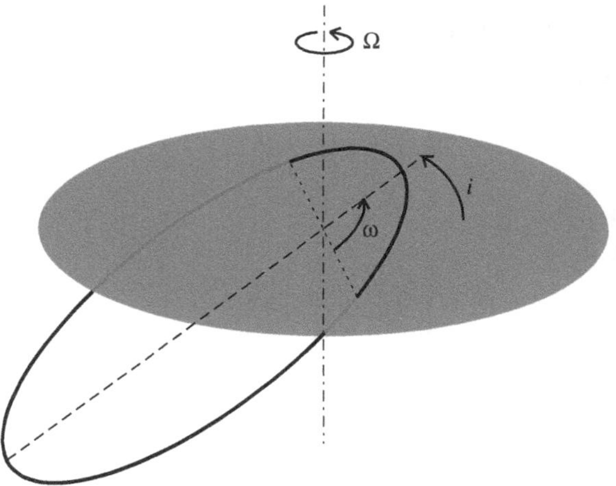

Fig. 1.1 The Keplerian angles: inclination i, longitude of the ascending node Ω and argument of pericentre ω. The filled *grey ellipse* denotes the reference plane

of five constants, the so-called Keplerian elements: semi-major axis a, eccentricity e, inclination i, longitude of the ascending node Ω, and argument of pericentre ω. The semi-major axis a and eccentricity e describe the geometrical shape of the ellipse. The three angles i, Ω and ω determine the orientation of the ellipse in space (see Fig. 1.1). In particular, inclination i and longitude of the ascending node Ω define the orientation of the orbital plane, i.e. the plane in which the ellipse lies. Argument of pericentre ω describes the orientation of the ellipse in this plane. The position of the star on the elliptic orbit is given by a non-constant element, the mean anomaly, ℓ. The orbital period, $T_{\rm orb}$, of the star reads as

$$T_{\rm orb} = 2\pi\sqrt{\frac{a^3}{GM_\bullet}}. \tag{1.1}$$

For future purposes, we find it useful to introduce the so-called eccentricity vector, $\boldsymbol{e}$, which is a vector that points to the pericentre of the orbit and whose magnitude equals the orbital eccentricity, $|\boldsymbol{e}| = e$,

$$\boldsymbol{e} \equiv \frac{\boldsymbol{v} \times \boldsymbol{L}}{GmM_\bullet} - \frac{\boldsymbol{r}}{|\boldsymbol{r}|}, \tag{1.2}$$

where $\boldsymbol{r}$, $\boldsymbol{v}$ and $\boldsymbol{L} \equiv \boldsymbol{r} \times m\boldsymbol{v}$ denote the position, velocity and angular momentum of the star, respectively. Furthermore, we define the directional angles of the eccentricity vector, Ω_e and i_e,

$$\cos\Omega_e \equiv -\frac{e_Y}{e_{XY}}, \tag{1.3}$$

$$\cos i_e \equiv \frac{e_Z}{|\boldsymbol{e}|}, \tag{1.4}$$

where e_X, e_Y and e_Z are components of the eccentricity vector in a Cartesian reference frame and $e_{XY} \equiv \sqrt{e_X^2 + e_Y^2} = |\boldsymbol{e}| \sin i_e$ represents its projection into the reference XY-plane.

1.2 Averaging Technique

Before we turn to description of some aspects of the dynamics in the perturbed Keplerian potential, it is useful to briefly describe a perturbation method that is commonly used in the celestial mechanics to simplify this problem, namely the so-called averaging technique (Morbidelli 2002; Bertotti et al. 2003).

For this purpose, we recall the star of mass m in the Keplerian potential of the SMBH of mass $M_\bullet$ from the previous section and write its Hamiltonian, $H_\bullet$, in terms of the classical position and velocity as

$$H_\bullet = \frac{mv^2}{2} - G\frac{mM_\bullet}{r}, \tag{1.5}$$

where $r \equiv |\boldsymbol{r}|$ and $v \equiv |\boldsymbol{v}|$. This Hamiltonian can be rewritten by means of the action-angle Delaunay variables which are defined by the Keplerian elements through relations

$$\begin{aligned} \mathcal{L} &= m\sqrt{GM_\bullet a}, & \ell &= \ell, \\ \mathcal{G} &= \mathcal{L}\sqrt{1-e^2}, & g &= \omega, \\ \mathcal{H} &= \mathcal{G}\cos i, & h &= \Omega, \end{aligned} \tag{1.6}$$

as

$$H_\bullet^{\mathrm{D}}(\mathcal{L}) = -\frac{G^2M_\bullet^2m^3}{2\mathcal{L}^2}. \tag{1.7}$$

Since this Hamiltonian depends only on $\mathcal{L}$, all the Delaunay variables except for ℓ are integrals of motion and, therefore, the Hamiltonian equations are solved trivially: $\mathrm{d}\ell/\mathrm{d}t = \mathrm{const.}$

Given the Keplerian potential of the SMBH is perturbed, however, the corresponding Hamiltonian may depend upon more Delaunay variables and the solution of the Hamiltonian equations may be much more complicated. Nevertheless, if the perturbation takes the form of an explicitly time-independent function $H_\epsilon^{\mathrm{D}}(\mathcal{L}, \mathcal{G}, \mathcal{H}, \ell, g, h)$ and the Hamiltonian of the system can be written as

$$H^{\mathrm{D}}(\mathcal{L}, \mathcal{G}, \mathcal{H}, \ell, g, h) = H_\bullet^{\mathrm{D}}(\mathcal{L}) + \epsilon H_\epsilon^{\mathrm{D}}(\mathcal{L}, \mathcal{G}, \mathcal{H}, \ell, g, h), \tag{1.8}$$

where ϵ is a small parameter, it is possible to find a transformation close to identity to a new set of variables in which the Hamiltonian is given by

$$H^{\mathrm{D}'}(\mathcal{L}', \mathcal{G}', \mathcal{H}', \ell', g', h') = H_{\bullet}^{\mathrm{D}}(\mathcal{L}') + \epsilon \overline{H}_1^{\mathrm{D}}(\mathcal{L}', \mathcal{G}', \mathcal{H}', -, g', h') + \epsilon^2 H_2^{\mathrm{D}'}(\mathcal{L}', \mathcal{G}', \mathcal{H}', \ell', g', h'). \tag{1.9}$$

In this formula, function $\overline{H}_1^{\mathrm{D}}(\mathcal{L}', \mathcal{G}', \mathcal{H}', -, g', h')$, which represents the averaged value of the perturbing component of the Hamiltonian (1.8) over the mean anomaly of the star ℓ, does not depend on the Delaunay variable ℓ' anymore. Hence, if we neglect the term of order ϵ^2, the Hamiltonian (1.9) does not depend on ℓ' and, therefore, $\mathcal{L}'$ and thus also $H_{\bullet}^{\mathrm{D}}(\mathcal{L}')$ are integrals of motion of the order of ϵ. In other words, the transformed semi-major axis a' of the orbit of the star as well as the corresponding transformed Keplerian energy in the potential of the SMBH is conserved to the order of ϵ. Similar to the function $\overline{H}_1^{\mathrm{D}}$, the transformed Keplerian elements (thus also the transformed Keplerian energy and other quantities) represent the averaged values of the osculating Keplerian elements over the mean anomaly of the star ℓ. The difference between the mean and osculating elements, however, is usually neglected as the corresponding transformation is close to an identity. We adopt this convention and do not distinguish between the two sets of orbital elements further on.

Since the Hamiltonian (1.9) is assumed to be explicitly time-independent, it is an integral of motion. As a result, also the function $\overline{H}_1^{\mathrm{D}}(\mathcal{L}', \mathcal{G}', \mathcal{H}', -, g', h')$ is conserved to the order of ϵ. Hence, in this order of approximation, the star is effectively replaced by an elliptical wire whose long-term evolution is determined by the orbit-averaged potential in the investigated system (see also Sect. 1.6).

1.3 Kozai-Lidov Mechanism

Among others, the averaging technique has been applied to many variants of the three-body problem. In particular, Kozai (1962) has investigated the restricted hierarchical circular case. In his setting (Sun–Jupiter–asteroid), the masses m_1, m_2 and m_3 of all bodies are well separated, $m_1 \gg m_2 \gg m_3$. Furthermore, the orbit of the perturbing body of mass m_2 around the dominating body of mass m_1 is assumed to be circular with radius R_2. Finally, the generally elliptical orbit of the body with the negligible mass m_3 lies within the orbit of the perturber, i.e. its semi-major axis, a_3, fulfils $a_3 < R_2$. In this case, averaging occurs over the mean anomalies of both the test and the perturbing body. Note that, for the long-term evolution of the orbit of the test body, the position of the perturbing body on its orbit is thus irrelevant, in other words, the perturbing body is equivalent to a circular, infinitesimally thin ring of constant linear density. Note that analogical problem has also been studied independently by Lidov (1962) for the case of the artificial satellites of the Earth whose motions are perturbed by the gravity of the Moon. Hence, as a tribute to both the pioneering researchers in this field, we refer to this type of configuration as the Kozai-Lidov problem.

For the purpose of our study, we consider the dominating body to be the SMBH of mass $M_\bullet$, the test body to be a star of mass m and the perturbing body to be a ring of mass M_r and radius R_r which can approximate gaseous torus or a secondary SMBH. In this notation, the quadrupole expansion of the mean perturbing potential energy reads as (Kozai 1962):

$$\overline{\mathcal{R}}_\mathrm{r} = -\frac{GmM_\mathrm{r}}{16R_\mathrm{r}}\left(\frac{a}{R_\mathrm{r}}\right)^2\left[\left(2+3e^2\right)\left(3\cos^2 i - 1\right) + 15e^2\sin^2 i\cos 2\omega\right], \quad (1.10)$$

where a, e, i and ω stand for the semi-major axis, eccentricity, inclination and argument of pericentre of the orbit of the star, respectively. As long as the ratio a/R_r is small enough, the higher terms in the expansion may be neglected.

As we can see, $\overline{R}_\mathrm{r}$ does not depend on the nodal longitude, Ω, of the stellar orbit (Delaunay variable $\mathcal{h}$), which simply reflects the axial symmetry of the orbit-averaged problem. Consequently, the z-component of the angular momentum of the star (Delaunay variable $\mathcal{H}$) is conserved, yielding a constant of motion

$$c \equiv \sqrt{1-e^2}\cos i \quad (1.11)$$

which is sometimes referred to as the Kozai integral.

The Kozai integral (1.11) enables us to eliminate the inclination in the averaged potential (1.10), leaving it to depend on the orbital eccentricity e and argument of pericentre ω, only. Hence, as the potential is also a conserved quantity in the orbit-averaged problem, the isolines $\overline{\mathcal{R}}_\mathrm{r} = C$ provide useful insights into the fundamental features of the dynamical evolution of both e and ω. This approach has been used by Kozai to discover two modes of topology of these isolines. When $c > \sqrt{3/5}$, the $\overline{\mathcal{R}}_\mathrm{r} = C$ isolines are simple ovals about the origin which is the only fixed point in the problem (see the left panel of Fig. 1.2). In this mode, eccentricity of the orbit does not vary much and the argument of pericentre rotates in interval $\langle 0, 2\pi\rangle$. For $c \leq \sqrt{3/5}$, the topology of the isolines becomes more complicated (see the right panel of Fig. 1.2). In particular, it contains two qualitatively different regions that are separated by a separatrix curve emerging from the origin, and also two new fixed points at non-zero eccentricity and argument of pericentre $\omega = \pi/2$ and $3\pi/2$. In the region in which the new two fixed points are found, the orbital argument of pericentre only oscillates in a rather narrow interval around the values that correspond to the new fixed points. For this reason, we shall refer to this region as the librational region, hereafter. In the second region, which shall be further denoted as the rotational region (outer rotational region; see the end of this section), the argument of pericentre rotates in the whole interval $\langle 0, 2\pi\rangle$.

The latter mode of the topology of the $\overline{\mathcal{R}}_\mathrm{r} = C$ isolines (right panel of Fig. 1.2) occurs whenever the initial inclination of the orbit is larger than $\approx 39.2°$, which is

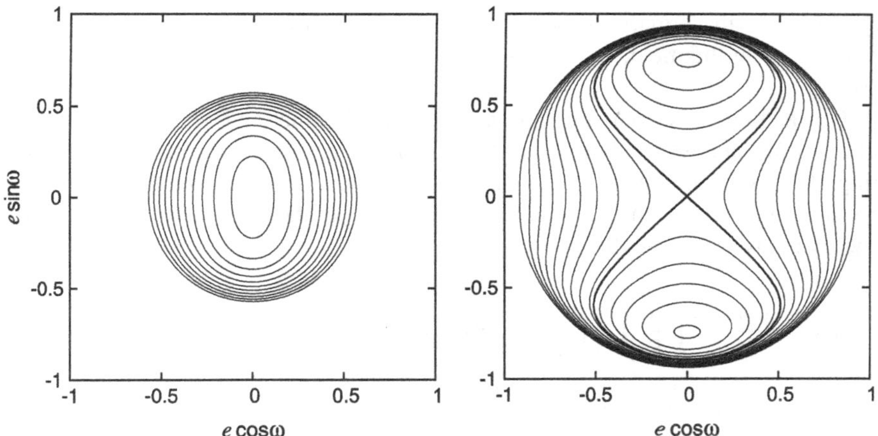

Fig. 1.2 Isolines of the conserved averaged potential of the ring, $\overline{\mathcal{R}}_{\rm r} = C$, from Eq. (1.10) for two different values of the Kozai integral of motion: $c = \cos(35°)$ in the *left panel* and $c = \cos(70°)$ in the *right panel* which correspond to 35° and 70° inclination circular orbit, respectively. The semi-major axis of the orbit is set to $a = 0.06\,R_{\rm r}$ and the mass of the ring to $M_{\rm r} = 0.3\,M_\bullet$. The origin $e = 0$ is a stationary point of the problem and in the first case is also stable, while in the second case it becomes unstable. The thick isoline in the *right panel* is the separatrix between two different regimes of eccentricity and pericentre evolution

sometimes called the Kozai limit. For such initial conditions, the orbit oscillates (in both the librational and rotational regions) between two extreme states due to conservation of c—either the orbit is highly inclined with low eccentricity or its eccentricity is high but the orbit lies nearly in the plane of the ring. These oscillations are commonly called the Kozai oscillations and the entire phenomenon is sometimes referred to as the Kozai-Lidov cycles (Kozai 1962; Lidov 1962).

In the quadrupole approximation, the Hamiltonian equations with the averaged potential energy (1.10) yield the following equations for the evolution of the Keplerian elements of the stellar orbit (Kiseleva et al. 1998; Kinoshita and Nakai 1999):

$$
\begin{aligned}
T_{\rm KL}\sqrt{1-e^2}\,\frac{{\rm d}e}{{\rm d}t} &= \frac{15}{8}\,e\left(1-e^2\right)\sin 2\omega \sin^2 i,\\
T_{\rm KL}\sqrt{1-e^2}\,\frac{{\rm d}i}{{\rm d}t} &= -\frac{15}{8}\,e^2 \sin 2\omega \sin i \cos i,\\
T_{\rm KL}\sqrt{1-e^2}\,\frac{{\rm d}\omega}{{\rm d}t} &= \frac{3}{4}\,\{2-2e^2+5\sin^2\omega\left[e^2-\sin^2 i\right]\},\\
T_{\rm KL}\sqrt{1-e^2}\,\frac{{\rm d}\Omega}{{\rm d}t} &= -\frac{3}{4}\,\cos i\left[1+4e^2-5e^2\cos^2\omega\right],
\end{aligned}
\tag{1.12}
$$

where

$$
T_{\rm KL} \equiv \frac{M_\bullet}{M_{\rm r}}\,\frac{R_{\rm r}^3}{a\sqrt{GM_\bullet a}} = \frac{1}{2\pi}\,\frac{M_\bullet}{M_{\rm r}}\left(\frac{R_{\rm r}}{a}\right)^3 T_{\rm orb} \tag{1.13}
$$

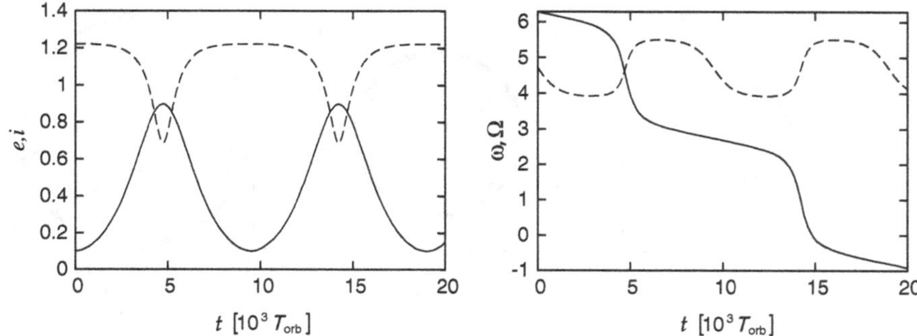

Fig. 1.3 Sample solution of Eqs. (1.12). Eccentricity e (*solid line*) and inclination i (*dashed line*) are shown in the *left panel*. The *right panel* displays the evolution of the nodal longitude Ω (*solid line*) and argument of pericentre ω (*dashed line*). Initial values $e_0 = 0.1$, $i_0 = 70°$, $\Omega_0 = 2\pi$ and $\omega_0 = 3\pi/2$ correspond to the upper librational lobe in the Kozai-Lidov diagrams (see the *right panel* of Fig. 1.2). Time is given in orbital periods. Mass of the ring is set to $M_r = 0.3\,M_\bullet$ and the semi-major axis of the orbit to $a = 0.06\,R_r$

is the characteristic timescale of the Kozai-Lidov cycles. A sample solution of these equations is shown in Fig. 1.3.

The Kozai-Lidov mechanism also occurs if the source of the axisymmetric perturbation is not an infinitesimally thin ring but a razor-thin disc. Since the conclusions for this configuration are similar to those discussed above for the ring-like perturber, let us only briefly mention a qualitative difference in the topology of the isolines for the averaged disc-like potential. As has been shown by Vokrouhlický and Karas (1998) and numerically tested in Šubr and Karas (2005), in such a case, the isolines contain, for low enough values of c (which is also an integral of motion in the case of the disc-like perturber), a third region—the inner rotational region (see Fig. 1.4). In this region, the Kozai oscillations of eccentricity and inclination are suppressed and the argument of pericentre rotates in the whole interval $\langle 0, 2\pi \rangle$ on the Kozai-Lidov timescale (1.13) in which the mass, M_r, and radius, R_r, of the ring are replaced by the mass, M_d, and characteristic radius, R_d, of the disc. The characteristic features of the two original regions (the librational and the outer rotational) remain qualitatively the same as in the case of the ring perturbation.

Finally, let us mention that the Kozai-Lidov dynamics has also been studied for other variants of the three-body problem. Ziglin (1975) has investigated a setting complementary to the classical Kozai-Lidov problem, assuming that the orbit of the test body lies outside the circular orbit of the perturber. The non-restricted version of the circular three-body problem has been fully described by (Lidov and Ziglin 1976; some aspects of the problem have already been discussed earlier; see the references in the paper). Most recently, Katz et al. (2011) have focused on the case when the orbit of the perturber is eccentric.

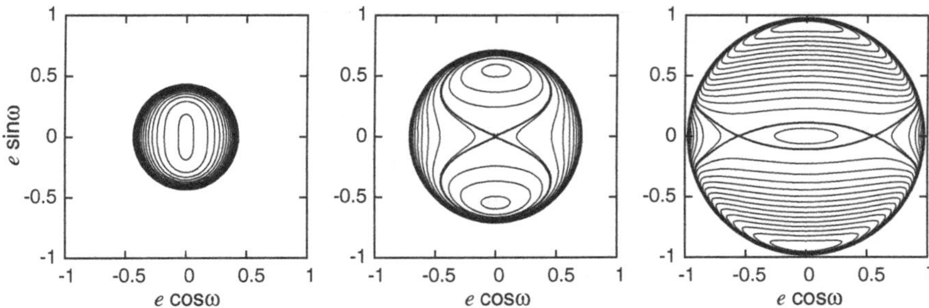

Fig. 1.4 Isolines of the conserved averaged potential in the case of a disc-like perturber. The value of the Kozai integral of motion decreases from left to right: $c = 0.9, 0.7$ and 0.2. In the *right panel*, we see the inner rotational region in which the initially circular orbit is stable. Mass of the disc is set to $M_\mathrm{d} = 0.01\,M_\bullet$ and the semi-major axis of the orbit to $a = 0.49\,R_\mathrm{d}$. The disc has constant surface density and its characteristic radius R_d is identified with its outer radius. The figure is redrawn from Šubr and Karas (2005)

1.4 Suppressed Kozai-Lidov Mechanism

In the previous section, we have investigated the Kozai-Lidov mechanism as it occurs in the case when the dominating Keplerian potential of the SMBH is perturbed only by an axisymmetric ring (disc). Here, we focus on the case when, besides the axisymmetric perturbation, an additional spherically symmetric potential is included, e.g. due to an extended star cluster.

We start with an assumption about the potential energy of a star of mass m in the spherical cluster. Considering a general power-law radial density profile of the cluster, $\rho\,(r) \propto r^{-\alpha}$ with $\alpha \neq 2$, we have the potential energy

$$\mathcal{R}_\mathrm{c} = -\frac{GmM_\mathrm{c}}{\beta R_\mathrm{r}}\left(\frac{r}{R_\mathrm{r}}\right)^{\beta}, \tag{1.14}$$

where $\beta = 2 - \alpha \neq 0$ and the mass of the cluster within the radius of the ring R_r is denoted M_c. According to the averaging technique, we shall integrate the potential energy (1.14) over one revolution about the centre with respect to the mean anomaly ℓ,

$$\overline{\mathcal{R}}_\mathrm{c} \equiv \frac{1}{2\pi}\int_{-\pi}^{\pi} \mathrm{d}\ell\ \mathcal{R}_\mathrm{c}, \tag{1.15}$$

which yields

$$\overline{\mathcal{R}}_\mathrm{c} = -\frac{1}{2\pi}\frac{GmM_\mathrm{c}}{\beta R_\mathrm{r}}\left(\frac{a}{R_\mathrm{r}}\right)^{\beta}\int_{-\pi}^{\pi} \mathrm{d}\ell\ \left(\frac{r}{a}\right)^{\beta}, \tag{1.16}$$

where $r(\ell)$ is given implicitly by relations $r = a\,(1 - e\cos u)$ and $u - e\sin u = \ell$. After some algebra, we obtain

$$\overline{\mathcal{R}}_{\mathrm{c}} = -\frac{GmM_{\mathrm{c}}}{\beta R_{\mathrm{r}}}\left(\frac{a}{R_{\mathrm{r}}}\right)^{\beta}\mathcal{J}\,(e,\beta)\,, \tag{1.17}$$

where

$$\mathcal{J}\,(e,\beta) \equiv \frac{1}{\pi}\int_0^{\pi} \mathrm{d}u\;(1 - e\cos u)^{1+\beta} = 1 + \sum_{n\geq 1} a_n e^{2n}, \tag{1.18}$$

with the coefficients obtained by recurrence

$$\frac{a_{n+1}}{a_n} = \left[1 - \frac{3+\beta}{2(n+1)}\right]\left[1 - \frac{2+\beta}{2(n+1)}\right] \tag{1.19}$$

and an initial value $a_1 = \beta\,(1+\beta)\,/4$. For the purpose of our study, we further set $\beta = 1/4$ which corresponds to the equilibrium model worked out by Bahcall and Wolf (1976,1977, see also Sect. 1.6) of this thesis.

Having discussed our assumptions about the spherical cluster, we now consider the combined effect of the ring and the cluster potentials on the long-term evolution of the stellar orbit. The total, orbit-averaged potential

$$\overline{\mathcal{R}} = \overline{\mathcal{R}}_{\mathrm{c}} + \overline{\mathcal{R}}_{\mathrm{r}} \tag{1.20}$$

still obeys axial symmetry, being independent on the nodal longitude. The picture, however, may be modified with respect to the case of solely ring-like perturbation. Considering the spherical cluster whose potential is approximated with (1.17), we find that the two types of topologies of the $\overline{\mathcal{R}} = C$ isolines persist (see Fig. 1.5) but the onset of the Kozai-Lidov oscillations of eccentricity and inclination of the orbit now depends on two parameters, namely c and $\mu \equiv M_{\mathrm{c}}/M_{\mathrm{r}}$. A non-zero mass of the cluster stabilizes small eccentricity evolution and the critical angle is pushed to larger values. For large enough μ, the stability of the circular orbit is guaranteed for arbitrary value of c and hence orbits of an arbitrary inclination with respect to the plane of the ring. This is because the effects of the cluster potential make the argument of pericentre circulate fast enough (significantly faster than the Kozai timescale), thus preventing secular changes in the eccentricity. An initially near-circular orbit maintains a very small value of e showing only small-amplitude oscillations. Figure 1.6 shows critical inclination values, for which the circular orbit becomes necessarily unstable as a function of μ and a/R_{r} parameters (note the later factorizes out from the analysis when $\mu = 0$). Importantly, there is a correlation between μ and a/R_{r} below which circular orbits of an arbitrary inclination are stable; for instance, data in Fig. 1.6 indicate that for $\mu = 0.1$ any circular orbit with $a \lesssim 0.12\,R_{\mathrm{r}}$ is stable.

Recalling Eqs. (1.12) for the evolution of the orbital elements, the impact of the spherical potential can be described by an additional term on the right-hand side of

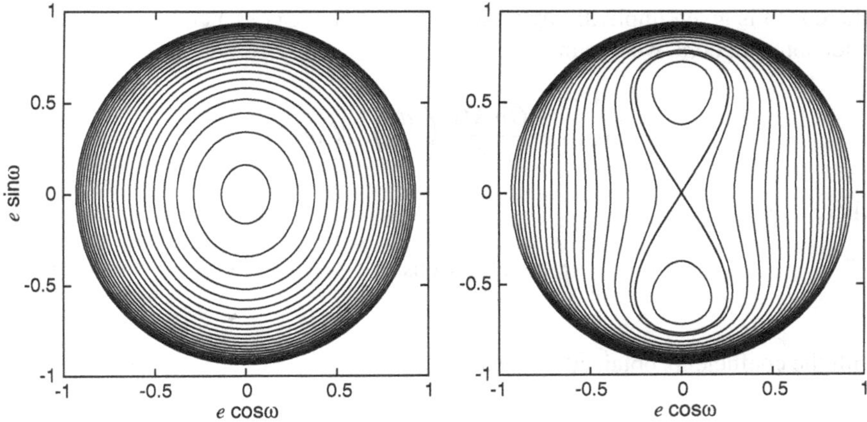

Fig. 1.5 Isolines of the conserved potential function $\overline{\mathcal{R}} = C$ from Eq. (1.20) for two different values of the mass ratio $\mu = M_c/M_r$: 0.1 and 0.01 in the *left* and *right panel*, respectively. The Kozai integral value is $c = \cos(70°)$, corresponding to 70° inclination circular orbit. Mass of the ring is set to $M_r = 0.3\ M_\bullet$ and the semi-major axis of the orbit to $a = 0.06\ R_r$. In the *left panel*, we see that sufficiently massive spherical cluster suppresses the Kozai-Lidov oscillations of eccentricity and inclination. The thick isoline in the *right panel* is a separatrix between two different regimes of eccentricity and pericentre evolution

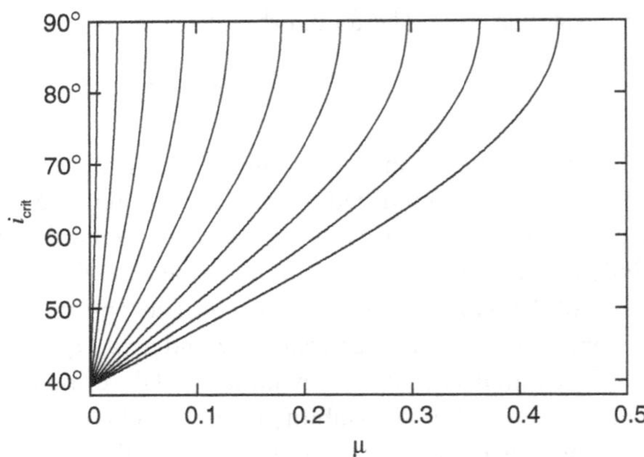

Fig. 1.6 Individual lines show a critical inclination (ordinate) at which the high-amplitude Kozai-Lidov oscillations onset for the given value of mass ratio $\mu = M_c/M_r$ (abscissa) for different values of orbital semi-major axis a ranging from $0.03\ R_r$ (*left*) to $0.3\ R_r$ (*right*) with the step of $0.03\ R_r$. When $\mu = 0$, the critical angle is ≈39.2° ('the Kozai limit') independently from a

the equation for argument of pericentre (Ivanov et al. 2005)

$$\left(\frac{d\omega}{dt}\right)_c = -K\frac{M_c\,(a)}{M_\bullet}\frac{\sqrt{1-e^2}}{T_{orb}}, \tag{1.21}$$

where $M_c(a)$ is the mass of the cluster enclosed with radius a and K is a dimensionless constant of order unity. The characteristic timescale, $T_c(a)$, of the pericentre rotation for the orbit with semi-major axis a then reads as (Ivanov et al. 2005; Merritt et al. 2011)

$$T_c(a) = k \frac{M_\bullet}{M_c(a)} T_{orb}(a) , \quad (1.22)$$

where k is a dimensionless constant of order unity.

1.5 Two-Body Relaxation

So far, we have been focusing on the dynamical effects of various mean potentials that can be present in general N-body stellar systems. We have seen that the motions of the individual stars in such potentials are deterministic and characterized by a set of integrals of motion. Namely, since the mean potential does not depend, by definition, on time, the total energy of each individual star in the system is conserved. Furthermore, depending on the spatial symmetries of the mean potential, also the angular momentum of the individual stars (or at least some of its components) might be conserved. However, as soon as the fluctuating component of the total potential in the system is present, the exact conservation of these integrals is being continuously disturbed. In particular, due to the either occasional or systematic stronger interaction of some of the stars in the system, both energy and angular momentum are exchanged among the individual orbits throughout the whole system, which leads to a slow evolution of its large-scale structure. This complex process, which eventually forces the stellar system to 'forget' its initial state, is commonly referred to as relaxation and the characteristic timescale on which this 'loss of memory' occurs is called the relaxation time. In the following, we shall review some of the basic features of relaxation which are important for our further research.

One of the most fundamental forms of relaxation is called two-body relaxation. In this case, the responsible part of the fluctuating component of the total potential corresponds to the short-term close two-body encounters that occur from time to time among the individual stars in the stellar system. Here, the term 'encounter' does not denote the physical contact between the stars (which we would refer to as 'collision'), it merely describes the situation in which the mutual interaction of the stars temporarily dominates, due to their spatial proximity over the effect of the mean potential of the system. As a result, during every such encounter, the orbits of the interacting stars are slightly modified as the stars exchange a small amount of both energy and angular momentum. A large number of these encounters thus lead to a global energy and angular momentum transfer throughout the whole system.

The rate of this transfer, i.e. the rate at which the stellar system relaxes from its initial state, can be estimated by means of evaluating the encounters for some of its typical stars. As the star travels through the system, the configuration of the surrounding stars fluctuates on timescale T_f, which depends on the stellar density and velocity dispersion in the system. During this interval, the star receives a typical impulse δv that changes its velocity v. Since the impulses in successive intervals of length T_f are uncorrelated, the total impulse Δv after time t evolves in a random walk fashion:

$$(\Delta v)^2 \sim (\delta v)^2 (t/T_{\rm f}). \tag{1.23}$$

Hence, we can now define the characteristic timescale, $T_{\rm tb}$, for two-body relaxation more quantitatively as the time that it takes to change the square of the original velocity of the star by order itself, $(\Delta v)^2 \sim v^2 \sim (\delta v)^2 (T_{\rm tb}/T_{\rm f})$. Inserting this relation back into (1.23), we thus obtain $(\Delta v)^2 \sim v^2 (t/T_{\rm tb})$.

As the total energy, E, of the star is, according to the virial theorem, proportional to the square of its velocity, $E \sim v^2$, the fluctuating component of the potential changes the energy at rate

$$\frac{\Delta E}{E} \sim \left(\frac{t}{T_{\rm tb}}\right)^{1/2}. \tag{1.24}$$

Similarly, the magnitude of the angular momentum of the star, $L \equiv |\boldsymbol{L}| \sim Rv$, diffuses at rate

$$\frac{\Delta L}{L_{\rm circ}} \sim \left(\frac{t}{T_{\rm tb}}\right)^{1/2}, \tag{1.25}$$

where $L_{\rm circ}$ denotes the maximum possible angular momentum of the star at the given energy level, i.e. in the case when its orbit is circular.

According to Binney and Tremaine (2008, Eq. (7.106)), the characteristic timescale, $T_{\rm tb}$, for a population of identical stars of mass m whose velocity distribution is Maxwellian with dispersion σ reads as

$$T_{\rm tb} \sim \frac{\sigma^3}{G^2 m \rho \ln \Lambda}, \tag{1.26}$$

where ρ is the spatial density of the stars, $\ln \Lambda$ represents the Coulomb logarithm and G denotes the gravitational constant. If the individual stars have different masses, the factor m should be replaced by the effective mass, $m_{\rm eff}$, which is defined as the ratio of the mean square mass to the mean mass of the stars in the investigated system, $m_{\rm eff} \equiv \int m^2 {\rm d}N(m) \,/ \int m {\rm d}N(m)$.

Under the assumption of virial equilibrium, the general expression (1.26) can be further simplified in order to obtain an order-of-magnitude estimate provided the population consists of N stars that are enclosed within a spherical volume of radius R. For such a stellar system, the velocity dispersion can be estimated as $\sigma^2 \sim v^2 \sim GNm/R$. Hence, estimating the stellar density in the system by its typical value, $\rho \sim Nm/R^3$, we find

$$T_{\rm tb} \sim \frac{N}{\ln \Lambda} T_{\rm cross}, \tag{1.27}$$

where the Coulomb logarithm approximately fulfils $\ln \Lambda \sim \ln N$ and $T_{\rm cross}$ is the crossing time, which is defined as the time it typically takes for a star to pass once through the whole system,

$$T_{\rm cross} \equiv \frac{R}{v} \sim \left(\frac{R^3}{GNm}\right)^{1/2}. \tag{1.28}$$

Using the simple estimate (1.27), we can immediately identify the astrophysical systems whose dynamical state must have already been affected by the two-body relaxation as their two-body relaxation time is shorter than or comparable to their age. Among these, we can name globular star clusters which contain $\sim 10^{4-6}$ stars, yielding $T_{\rm tb} \sim 10^{3-4}\ T_{\rm cross}$, and are $\sim 10^4\ T_{\rm cross}$ old. On the other hand, for the dynamical evolution of the systems whose two-body relaxation time is much larger than their age, the short-term close two-body encounters are not important. Here, galaxies would serve as an example since they consist of $\sim 10^{11}$ stars, their relaxation time reaches $T_{\rm tb} \sim 10^9\ T_{\rm cross}$ and their age is of order $\sim 10^2\ T_{\rm cross}$.

1.6 Central Mass Dominated Relaxation

In the following, we focus on specific features of relaxation in systems whose gravitational potential is dominated by the central SMBH. For this purpose, we consider a spherically symmetric cluster of N stars of mass m on orbits around the SMBH of mass $M_\bullet \gg Nm$. The potential of the SMBH is assumed to be Keplerian and, therefore, the mean potential within the cluster is near-Keplerian. If the fluctuating component of the potential were not present, the individual orbits would be Kepler ellipses, undergoing a negative pericentre shift due to the mean potential of the stars on a timescale (1.22), which is much longer than the orbital period of the stars, $T_{\rm c} \gg T_{\rm orb}$.

An order-of-magnitude estimate of the two-body relaxation time in such a star cluster can be obtained as follows (cf. the relaxation time (1.27)). Since the stellar motions in the cluster are dominated by the Keplerian potential of the SMBH, the Jeans equations dictate $\sigma^2 \sim v^2 \sim GM_\bullet/R$, where R is the typical radius of the orbits in the cluster. Inserting this velocity dispersion to the general formula (1.26) and estimating the stellar density in the cluster by its typical value, $\rho \sim Nm/R^3$, yields (see also Rauch and Tremaine 1996, Eq. (51))

$$T_{\rm tb} \sim \frac{M_\bullet^2}{m^2 N \ln N} T_{\rm orb}, \tag{1.29}$$

where we have applied the approximate relation $\ln\Lambda \sim \ln N$. Given the stars in the cluster are not equal-mass, the factor m in the relaxation time should be replaced by the effective mass $m_{\rm eff}$.

As an illustration, let us calculate the relaxation time for a star cluster with parameters that are expected to be quite common for clusters in galactic nuclei, including the one of our own Galaxy: $M_\bullet \sim 10^6\,M_\odot$, $N \sim 10^5$, $m \sim 1\,M_\odot$, $R \sim 1$ pc. In such a case, the typical orbital period of the stars is of order $\sim 10^4$yr and the relaxation time reaches $\sim 10^9$ yr. Since the estimated age of the nuclear star clusters is of the same order of magnitude, we see that these clusters are likely to be affected by the two-body relaxation.

The evolution of spherically symmetric star clusters harbouring the central SMBH due to two-body relaxation has been studied by various authors (Peebles 1972a,b; Bahcall and Wolf 1976, 1977). For the purpose of this thesis, let us only briefly mention that it leads to a steady-state distribution of the stars which is described by the energy distribution function $f(E) \propto E^{1/4}$ (Bahcall and Wolf 1976) if the cluster accommodates several assumptions, such as equal masses of the stars. In terms of the stellar density, $\rho(r)$, this distribution corresponds to

$$\rho(r) \propto r^{-7/4} \tag{1.30}$$

which is usually called the Bahcall-Wolf radial density profile.

The stellar motions in the near-Keplerian mean potential of the central mass dominated star cluster are regular. Consequently, the individual stars in the cluster can interact coherently on a timescale that is much longer than their orbital period, which leads to a much more efficient exchange of the angular momentum among the stellar orbits. As a result, the cluster undergoes an enhanced relaxation of angular momentum which is usually referred to as resonant relaxation (Rauch and Tremaine 1996).

In order to understand the basic principles behind this dynamical process, let us first imagine averaging of the stellar trajectories in the cluster over an interval that is $\gg T_{\mathrm{orb}}$ but $\ll T_{\mathrm{c}}$ given by (1.22). On this timescale, the individual stars in the cluster are represented by fixed eccentric wires. The mass of each wire equals the mass of the original star, its shape corresponds to the shape of the original stellar orbit and the linear density of the wire varies in accord with the non-uniform motion of the star on this orbit. The long-term interactions of the stars in the cluster can then be described in terms of the torques exerted among these wires. The gravitational potential generated by the wires is roughly stationary, similar to the case of the previously discussed Kozai-Lidov problem (see Sects. 1.2 and 1.3). Hence, the torques among the wires do not lead to any exchange of energy, $\Delta E = 0$. The angular momentum distribution in the system, however, is affected significantly. This process is usually called the scalar resonant relaxation and it changes both the magnitude and direction of the angular momenta of the wires.

Due to the pericentre shift in the mean potential of the cluster, the configuration of the wires fluctuates on timescale T_{c} (see Eq. (1.22)) and, consequently, so do the mutual torques among them. During an interval of length T_{c}, the torques change (roughly coherently) the magnitude of the angular momentum of a typical wire by some amount δL. On timescales $t \gg T_{\mathrm{c}}$, the total angular momentum change, ΔL, of the wire follows a random walk with increments δL and, therefore, $(\Delta L)^2 \sim (\delta L)^2 (t/T_{\mathrm{c}})$. In analogy with the two-body relaxation time, we can define the scalar resonant relaxation time, T_{sr}, as the time that it takes to change the square of the original angular momentum by order L_{circ}^2: $(\Delta L)^2 \sim L_{\mathrm{circ}}^2 \sim (\delta L)^2 (T_{\mathrm{sr}}/T_{\mathrm{c}})$. Hence, we obtain (cf. Eq. (1.25))

$$\frac{\Delta L}{L_{\rm circ}} \sim \left(\frac{t}{T_{\rm sr}}\right)^{1/2}, \qquad t \gg T_{\rm c}. \tag{1.31}$$

As has been first derived by Rauch and Tremaine (1996) and later discussed by Hopman and Alexander (2006) and Eilon et al. (2009), the scalar resonant relaxation time is given by

$$T_{\rm sr} \sim \frac{M_\bullet}{m} T_{\rm orb}. \tag{1.32}$$

In the case of non-equal-mass stars, the factor m is replaced by the effective mass $m_{\rm eff}$. Let us emphasize here that the scalar resonant relaxation time is shorter than the two-body relaxation time (1.29) by a factor $(mN \ln N)/M_\bullet$ which is $\ll 1$ in the case of near-Keplerian potential ($M_\bullet \gg Nm$). This is a direct consequence of the fact that the stars can coherently interact on a much longer timescale (in comparison to their short-term two-body encounters) and, therefore, exchange their angular momentum much more efficiently. In this context, let us also note that Rauch and Tremaine (1996) have pointed out that the Bahcall-Wolf radial density profile (1.30) may not correctly describe the structure of the innermost parts of the relaxed spherically symmetric star cluster around the central SMBH as this profile has been derived considering two-body relaxation only.

Another form of resonant relaxation occurs in general spherical potentials or even in axisymmetric potentials that are nearly spherical. This process, which is known as vector resonant relaxation, affects only the direction of the angular momentum (or its z-component which is the integral of motion in the case of axisymmetric potentials), leaving its magnitude untouched. Similar to the case of the scalar resonant relaxation, the diffusion of the angular momentum due to the vector resonant relaxation is also caused by the torques that arise from the averaged mass distribution in the cluster. In this case, however, the averaging is done over a longer timescale $T_{\rm c} \ll t \ll T_{\rm prec}$, where $T_{\rm prec}$ denotes the characteristic timescale of precession of the angular momentum of the orbits in the cluster that is caused by the fluctuating component of the spherical potential (or by the axisymmetric potential). The individual stars are, on this timescale, smeared into axisymmetric annuli whose inner and outer radius correspond to the pericentre and apocentre of the original orbits, respectively. Again, the potential generated by the annuli is roughly stationary and, therefore, the torques do not affect their energy, $\Delta E = 0$. Contrary to the case of the scalar resonant relaxation, however, the torques are perpendicular to the normals of the annuli and, as such, do not lead to any change of the magnitude of angular momentum, $\Delta L = 0$.

The configuration of the annuli fluctuates on time-scale $T_{\rm prec}$. In spherical potentials, the total change of the vector angular momentum, $|\Delta \boldsymbol{L}|$, of a typical annulus after time $t \gg T_{\rm prec}$ is, therefore, described by a random walk with a step-size $T_{\rm prec}$ and increments, $|\delta \boldsymbol{L}|$, accumulated in each step: $|\Delta \boldsymbol{L}|^2 \sim |\delta \boldsymbol{L}|^2 (t/T_{\rm prec})$. However, as has been shown by Rauch and Tremaine (1996), already $|\delta \boldsymbol{L}|/L_{\rm circ} \sim 1$. Hence, the characteristic timescale, $T_{\rm vr,sf}$, of the vector resonant relaxation in spherical potentials reads as

$$T_{\mathrm{vr,sf}} \sim T_{\mathrm{prec}} \sim \frac{N^{1/2}}{\mu} T_{\mathrm{orb}}, \tag{1.33}$$

where $\mu \equiv Nm/(M_\bullet + Nm)$. If the potential is near-Keplerian, we see that $T_{\mathrm{vr,sf}}$ is shorter than the scalar resonant relaxation time (1.32) by a factor of $N^{-1/2}$.

In the case of axisymmetric potentials, the period T_{prec} is given by the precession rate in these potentials. On timescales $t \gg T_{\mathrm{prec}}$, the total change in the z-component of the angular momentum, ΔL_z, of a typical annulus evolves in a random walk fashion with increments, δL_z, accumulated over T_{prec}: $|\Delta L_z|^2 \sim |\delta L_z|^2 (t/T_{\mathrm{prec}})$. Hence, if we define the vector resonant relaxation time, $T_{\mathrm{vr,axi}}$, as the time when the change reaches $|\Delta L_z|^2 \sim L_{\mathrm{circ}}^2 \sim |\delta L_z|^2 (T_{\mathrm{vr,axi}}/T_{\mathrm{prec}})$, we find

$$\frac{|\Delta L_z|}{L_{\mathrm{circ}}} \sim \left(\frac{t}{T_{\mathrm{vr,axi}}}\right)^{1/2}, \qquad t \gg T_{\mathrm{prec}}. \tag{1.34}$$

In this case, $T_{\mathrm{vr,axi}}$ fulfils (Rauch and Tremaine 1996)

$$T_{\mathrm{vr,axi}} \sim \frac{N}{\mu^2} \frac{T_{\mathrm{orb}}^2}{T_{\mathrm{prec}}} \tag{1.35}$$

which is shorter by a factor $(T_{\mathrm{orb}}/T_{\mathrm{prec}}) \ln N$ than the characteristic timescale of the two-body relaxation (1.29) if the potential is near-Keplerian.

1.7 Relaxation of Thin Stellar Discs

The third power of the velocity dispersion in the two-body relaxation time (1.26) indicates that the rate of this process strongly depends on the relative velocities of the stars in the system under consideration. Consequently, we may expect the two-body relaxation to be much more efficient in systems in which the stars are moving coherently, i.e. with small relative velocities, such as thin stellar discs in the dominating potential of the central SMBH.

If the disc is formed by N stars of mass m that are coherently orbiting the central SMBH of mass $M_\bullet \gg Nm$, the velocity dispersion, σ, can be written as $\sigma \sim e_{\mathrm{rms}} v_{\mathrm{K}}$ (e.g., Lissauer 1993), where $v_{\mathrm{K}} \sim (GM_\bullet/R)^{1/2}$ is the Keplerian velocity, R stands for the typical radius of the orbits in the disc and e_{rms} denotes their root mean square eccentricity. The stellar density, ρ, in the disc can be estimated by $\rho \sim Nm/hR^2$ in which h is the scale height of the disc that can be expressed as $h \sim \sigma R/v_{\mathrm{K}}$. Inserting these quantities into the two-body relaxation time (1.26) leads to (for details, see Stewart and Ida 2000; Alexander et al. 2007)

$$T_{\mathrm{tb,disc}} \sim \frac{M_\bullet^2 e_{\mathrm{rms}}^4}{m^2 N \ln \Lambda} T_{\mathrm{orb}}. \tag{1.36}$$

In this formula, Λ can be calculated as (Stewart and Ida 2000)

$$\Lambda \sim \frac{M_\bullet e_{\rm rms}^3}{m}. \tag{1.37}$$

Formula (1.37) holds if $\Lambda \gg 1$ (which is the usual case in stellar dynamics) and if the root mean square values of orbital eccentricity, $e_{\rm rms}$ and inclination, $i_{\rm rms}$, in the disc follow relation

$$e_{\rm rms} \approx 2 i_{\rm rms} \tag{1.38}$$

which has been shown to be a natural consequence of two-body relaxation of thin stellar discs (e.g., Ida et al. 1993).

Hence, we see that the two-body relaxation time (1.36) may indeed be very short if the orbits in the disc are near-circular. This is in agreement with the physical intuition since, due to low relative velocities of the stars in such a disc, the stars can interact on a longer timescale and, therefore, exchange a larger amount of both energy and angular momentum.

The magnitude of the angular momentum of a star from the disc is given by

$$L = L_{\rm circ}\sqrt{1-e^2}, \tag{1.39}$$

where

$$L_{\rm circ} = m\sqrt{GM_\bullet a}. \tag{1.40}$$

Differentiation of L thus yields

$$\Delta L = -L_{\rm circ}\frac{e}{\sqrt{1-e^2}}\Delta e + m\sqrt{GM_\bullet\left(1-e^2\right)}\frac{1}{2a^{1/2}}\Delta a, \tag{1.41}$$

which can be, for sufficiently small eccentricities, rewritten as

$$\Delta L \approx -L_{\rm circ} e \Delta e + m\sqrt{GM_\bullet}\frac{1}{2a^{1/2}}\Delta a. \tag{1.42}$$

Since the total energy, E, of the star is given by

$$E = -\frac{GmM_\bullet}{2a}, \tag{1.43}$$

we have

$$\Delta L \approx -L_{\rm circ}\left[\frac{\Delta\left(e^2\right)}{2} + \frac{\Delta E}{2E}\right]. \tag{1.44}$$

If we substitute relations (1.24) and (1.25) and average over the stars in the disc, we find that the two-body relaxation of thin stellar discs leads to an increase of the root

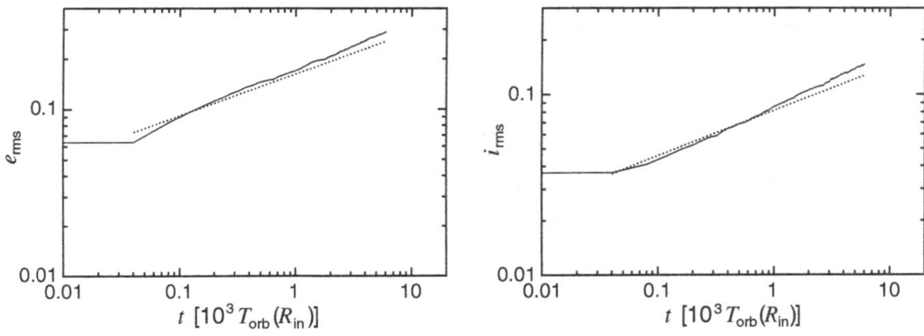

Fig. 1.7 Two-body relaxation of the stellar disc around the SMBH: evolution of the root mean square eccentricity e_{rms} (*left panel*) and inclination i_{rms} (*right panel*) of the orbits in the disc. Time is given in multiples of the orbital period that corresponds to the initial inner radius of the orbits in the disc, R_{in}. We see that the evolution of both mean elements roughly agrees with the theoretical $t^{1/4}$ dependence (*thin dotted lines*; see Eqs. (1.45) and (1.46)). The displayed results describe one realization of the model (see Table 2.1 for model parameters, model B0)

mean square eccentricity of the orbits by amount

$$\Delta e_{rms} \sim \left(\frac{t}{T_{tb,disc}}\right)^{1/4} . \quad (1.45)$$

Furthermore, due to relation (1.38), the root mean square inclination evolves in the same fashion as

$$\Delta i_{rms} \sim \left(\frac{t}{T_{tb,disc}}\right)^{1/4} . \quad (1.46)$$

Relations (1.45) and (1.46) may be straightforwardly tested by means of numerical N-body modelling. For this purpose, we use the publicly available N-body integration code NBODY6 (Aarseth 2003) and follow the orbital evolution of an initially thin stellar disc in the dominating Keplerian potential of the central SMBH. The Keplerian potential has been implemented into the original version of the code as an additional external potential. The stellar orbits in the disc are assumed to be initially circular with radii $R \in \langle R_{in}, 10R_{in}\rangle$ (for the remaining parameters of the model, see Table 2.1, model B0).

Evolution of the root mean square eccentricity, e_{rms}, and inclination, i_{rms}, of the orbits in the disc is shown in the left and right panel of Fig. 1.7, respectively. As we can see, both elements evolve in a rough accord with the theoretically predicted dependence $\sim t^{1/4}$ (thin dotted lines; Eqs. (1.45) and (1.46); see also Stewart and Ida 2000; Alexander et al. 2007; Cuadra et al. 2008). Figure 1.8 demonstrates the evolution of the orbital semi-major axes in the disc in terms of their initial (grey rectangle) and final distribution (boxes; $t \approx 10^4\, T_{orb}(R_{in})$). We observe that the evolved distribution differs significantly from the initial state.

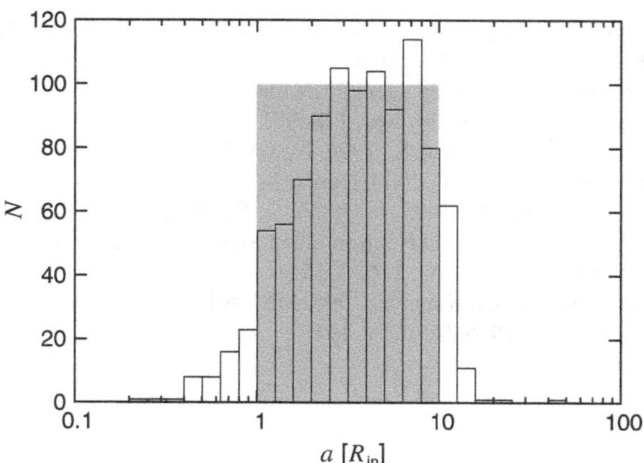

Fig. 1.8 Two-body relaxation of the stellar disc around the SMBH: comparison of the initial (*grey rectangle*) and final distribution (*boxes*; $t \approx 10^4\, T_{\rm orb}(R_{\rm in})$) of the orbital semi-major axes in the disc. We see that both distributions differ significantly. The displayed results describe one realization of the model (see Table 2.1 for the model parameters, model B0)

The orbital evolution of thin stellar discs may also be affected by resonant relaxation. However, as discussed in Kocsis and Tremaine (2011) in the context of the young stellar system in the Galactic Centre, the vector resonant relaxation of thin stellar discs is much less efficient than their two-body relaxation, unless they are embedded in an extended spherical star cluster. The determination of the scalar resonant relaxation rate for thin stellar discs that are not exposed to any perturbative sources of gravity is not straightforward and we refer the reader to the work of Tremaine (1998).

References

Aarseth S. J., 2003, Gravitational N-Body Simulations. Cambridge Univ. Press, Cambridge
Alexander R. D., Begelman M. C., Armitage P. J., 2007, ApJ, 654, 907
Bahcall J. N., Wolf R. A., 1976, ApJ, 209, 214
Bahcall J. N., Wolf R. A., 1977, ApJ, 216, 883
Bertotti B., Farinella P., Vokrouhlický D., 2003, Physics of the Solar System. Kluwer Academic Publishers, Dordrecht
Binney J., Tremaine S., 2008, Galactic Dynamics. Princeton Univ. Press, Princeton
Cuadra J., Armitage P. J., Alexander R. D., 2008, MNRAS, 388, L64
Eilon E., Kupi G., Alexander T., 2009, ApJ, 698, 641
Hopman C., Alexander T., 2006, ApJ, 645, 1152
Ida S., Kokubo E., Makino J., 1993, MNRAS, 263, 875
Ivanov P. B., Polnarev A. G., Saha P., 2005, MNRAS, 358, 1361
Katz B., Dong S., Malhotra R., 2011, PhRvL, 107, 181101
Kinoshita H., Nakai H., 1999, CeMDA, 75, 125
Kiseleva L. G., Eggleton P. P., Mikkola S., 1998, MNRAS, 300, 292
Kocsis B., Tremaine S., 2011, MNRAS, 412, 187
Kozai Y., 1962, AJ, 67, 591
Lidov M. L., 1962, Planet. Space Sci., 9, 719
Lidov M. L., Ziglin S. L., 1976, CeMec, 13, 471
Lissauer J. J., 1993, ARA&A, 31, 129

Merritt D., Alexander T., Mikkola S., Will C. M., 2011, PhRvD, 84, 044024
Morbidelli A., 2002, Modern Celestial Mechanics. Taylor & Francis, London
Peebles P. J. E., 1972a, Gen. Rel. Grav., 3, 63
Peebles P. J. E., 1972b, ApJ, 178, 371
Rauch K. P., Tremaine S., 1996, NewA, 1, 149
Stewart G. R., Ida S., 2000, Icarus, 143, 28
Šubr L., Karas V., 2005, in Hledík S., Stuchlík Z., eds, Proceedings of RAGtime 6/7: Workshops on Black Holes and Neutron Stars. Silesian University in Opava, Czech Republic, p. 281
Tremaine S., 1998, AJ, 116, 2015
Vokrouhlický D., Karas V., 1998, MNRAS, 298, 53
Ziglin S. L., 1975, SvAL, 1, 194

Chapter 2
Thin Disc Embedded in Extended Spherically Symmetric Cluster

In Sect. 1.7, we have investigated the orbital evolution of thin isolated stellar discs around the SMBH. There is statistically significant evidence that such structures, indeed, are present in galactic nuclei (Levin and Beloborodov 2003; Bender et al. 2005; Paumard et al. 2006; Bartko et al. 2009, 2010; Lauer et al. 2012). In this case, however, it is reasonable to expect the stellar disc to be embedded in an extended spherically symmetric relaxed star cluster that is centred on the SMBH. Hence, in this chapter, we focus on how the orbital evolution of the stellar discs is affected if we include the spherical perturbation.

2.1 The Cluster Modelled by Analytic Potential

In the first step, we emulate the gravity of the extended spherical star cluster by an analytic spherically symmetric potential, i.e. we neglect the fluctuating component of the total potential generated by the cluster. In order to test whether such a perturbation has an impact upon the relaxation of the disc, we consider the same stellar disc as in Sect. 1.7 and include the potential that corresponds to the spherically symmetric mass distribution described by the Bahcall-Wolf radial density profile (1.30) (see model B1 in Table 2.1 for the particular values of the model parameters). As this profile is expected to describe the structure of relaxed spherical star clusters that contain a massive central body (see Sect. 1.6), it represents a natural choice for modelling the mean potential of the cluster in our calculations. The strength of the potential is characterized by the mass, $M_c(R_{out})$, of the cluster enclosed within the outer radius of the disc, $R_{out} = 10\, R_{in}$, and we set $M_c(R_{out}) = 0.01\, M_\bullet$.

Figure 2.1 shows the evolution of the root-mean-square eccentricity, e_{rms}, and inclination, i_{rms}, of the orbits in the disc (solid lines). If we compare the results with those acquired for the isolated stellar disc (dashed lines; redrawn from Fig. 1.7; model B0), we cannot identify any significant difference, except for the initially larger values of e_{rms} due to the effectively larger mass enclosed within the orbits. Similar results can be obtained even for larger characteristic masses $M_c(R_{out})$, as long as the

J. Haas, *Symmetries and Dynamics of Star Clusters*,
Springer Theses, DOI: 10.1007/978-3-319-03650-2_2,
© Springer International Publishing Switzerland 2014

Table 2.1 Parameters of the numerical models described in Chaps. 1 and 2

	B0	B1	canon.	H	M	A	D
N-body disc	yes	yes	yes	yes	yes	yes	yes
N-body cluster	–	–	yes	yes	yes	yes	yes
Predef. cluster	–	yes	–	–	yes	–	–
Predef. ring	–	–	–	–	–	–	–
N_{d1}	600	600	600	600	600	600	600
m_{d1} $[10^{-6}\,M_\bullet]$	2.5	2.5	2.5	2.5	2.5	2.5	2.5
N_{d2}	300	300	300	300	300	300	300
m_{d2} $[10^{-6}\,M_\bullet]$	7.5	7.5	7.5	7.5	7.5	7.5	7.5
N_{d3}	100	100	100	100	100	100	100
m_{d3} $[10^{-6}\,M_\bullet]$	25	25	25	25	25	25	25
$a_{d,min}$–$a_{d,max}$ $[R_{in}]$	1–10	1–10	1–10	1–10	1–10	1–10	1–10
α_d	-1	-1	-1	-1	-1	-1	-1
e_d	gc	gc	gc	gc	gc	gc	gc
Δ [°]	2.5	2.5	2.5	2.5	2.5	2.5	2.5
N_c $[10^3]$	–	–	12.5	100	6.25	50	6.25
m_c $[10^{-6}\,M_\bullet]$	–	–	1.903	2.378	3.806	0.476	3.806
$a_{c,min}$–$a_{c,max}$ $[R_{in}]$	–	–	0.5–20	0.5–20	0.5–20	0.5–20	0.5–20
α_c	–	–	1/4	1/4	1/4	1/4	1/4
e_c	–	–	gc	gc	gc	gc	gc
$M_c(R_{out})$ $[10^{-2}\,M_\bullet]$	–	1	–	–	1	–	–
β_c	–	$-7/4$	–	–	$-7/4$	–	–
M_r $[10^{-1}\,M_\bullet]$	–	–	–	–	–	–	–
R_r $[R_{in}]$	–	–	–	–	–	–	–
i_r	–	–	–	–	–	–	–
Num. integrator	N	N	N	N	N	N	N

	K0a	K0b	K1a	K1b
N-body disc	–	–	yes	yes
N-body cluster	yes	yes	–	–
Predef. cluster	–	–	–	–
Predef. ring	yes	yes	yes	yes
N_{d1}	–	–	300	300
m_{d1} $[10^{-6}\,M_\bullet]$	–	–	tp	4
N_{d2}	–	–	–	–
m_{d2} $[10^{-6}\,M_\bullet]$	–	–	–	–
N_{d3}	–	–	–	–
m_{d3} $[10^{-6}\,M_\bullet]$	–	–	–	–
$a_{d,min}$–$a_{d,max}$ $[R_{in}]$	–	–	1–1.8	1–1.8
α_d	–	–	-1	-1
e_d	–	–	0–0.1	0–0.1
Δ [°]	–	–	2.5	2.5

(continued)

Table 2.1 (continued)

	K0a	K0b	K1a	K1b
N_c [10^3]	0.6	0.6	–	–
m_c [$10^{-6}\,M_\bullet$]	tp	25	–	–
$a_{c,min}$–$a_{c,max}$ [R_{in}]	1–3*	1–3*	–	–
α_c	1/4	1/4	–	–
e_c	0–0.1	0–0.1	–	–
$M_c(R_{out})$ [$10^{-2}\,M_\bullet$]	–	–	–	–
β_c	–	–	–	–
M_r [$10^{-1}\,M_\bullet$]	1	1	0.5	0.5
R_r [R_{in}]	12*	12*	6	6
i_r	–	–	$\pi/2$	$\pi/2$
Num. integrator	M	M	M	M

The first five rows give the model designation and included components in order: the N-body disc, the N-body cluster, the predefined analytic cluster and the predefined analytic ring. The subsequent rows describe the initial properties of the N-body disc: numbers (N_{d1}, N_{d2}, N_{d3}) and masses (m_{d1}, m_{d2}, m_{d3}) of the particles in the disc, interval ($a_{d,min}$, $a_{d\,max}$) and power-law index (α_d) of the initial distribution of the orbital semi-major axes in the disc, initial eccentricities of the orbits in the disc (e_d) and initial half-opening angle of the disc (Δ). The analogical quantities indexed by 'c' describe the initial properties of the N-body cluster. The characteristic mass of the predefined analytic cluster within the initial outer radius of the disc R_{out} and the power-law index of the corresponding radial density profile, $\rho \sim r^{\beta_c}$, are denoted $M_c(R_{out})$ and β_c, respectively. The predefined analytic ring is described by its mass M_r, radius R_r and inclination i_r with respect to the plane of the disc. The row 'Num. integrator' indicates which N-body integration code has been used for the particular model: NBODY6 (N; Aarseth 2003) and Mbody (M; Šubr 2006). The abbreviation 'gc' indicates that the orbits are constructed to be geometrically circular; 'tp' stands for 'test particles' with extremely low mass $2.5 \times 10^{-12}\,M_\bullet$. In the case of models K0a and K0b, the length unit is chosen to be the inner radius of the cluster since these models do not include any disc whose inner radius, R_{in}, represents the length unit in the remaining models (the affected values are marked by an asterisk)

potential of the cluster may be considered a perturbation to the dominating potential of the SMBH ($M_c(R_{out}) \lesssim 0.1\,M_\bullet$). In conclusion, we find that the gravitational perturbation in the form of an analytic spherically symmetric potential does not affect relaxation of the angular momentum in the disc.

Furthermore, comparison of the evolved distributions of the orbital semi-major axes displayed in Fig. 2.2 reveals that, within both models, the distributions are of the same shape. Hence, it appears that neither the energy transfer in the disc is affected if the analytic spherical cluster is included. The case when the cluster is modelled by a large number of gravitating, mutually interacting particles is described in the following section.

2.2 Coupling of the Disc and the Cluster

In order to investigate the orbital evolution of the thin stellar disc under the gravitational influence of the spherical cluster which is treated in the full N-body way, we introduce the following configuration:

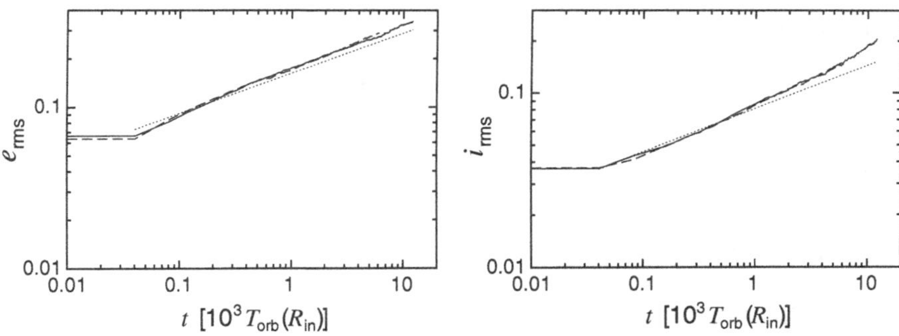

Fig. 2.1 Evolution of the root-mean-square eccentricity e_{rms} (*left panel*) and inclination i_{rms} (*right panel*) of the orbits in the stellar disc around the SMBH in a predefined analytic spherically symmetric potential (*solid lines*; model B1) in comparison to the previously discussed case of the isolated stellar disc (*dashed lines*; redrawn from Fig. 1.7; model B0).Time is given in multiples of the orbital period that corresponds to the initial inner radius of the orbits in the disc, R_{in}. The *thin-dotted lines* denote the theoretical $t^{1/4}$ dependence derived for the isolated stellar discs (see Eqs. (1.45) and (1.46)). The displayed results correspond to 1 realization of the models. In the case of model B1, the parameters are the same as in Fig. 1.7, except for $M_c(R_{out}) = 0.01\ M_\bullet$ (see also Table 2.1)

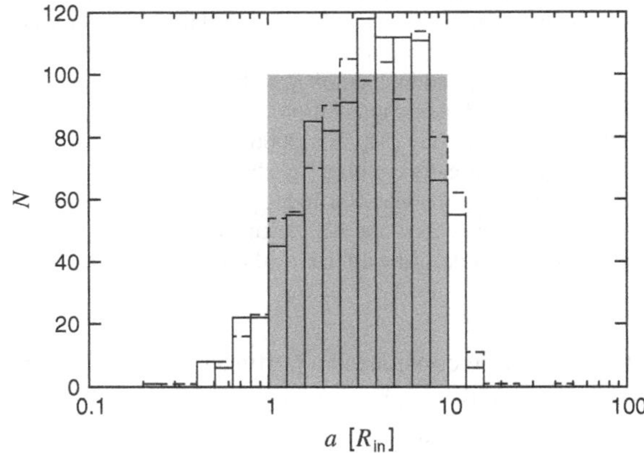

Fig. 2.2 Evolution of the distribution of the orbital semi-major axes in the stellar disc around the SMBH in a predefined analytic spherically symmetric potential (*solid boxes*; model B1) in comparison to the previously discussed case of the isolated stellar disc (*dashed boxes*; redrawn from Fig. 1.8; model B0). The initial distribution (*grey rectangle*) is the same in both cases. The evolved distributions describe the state at $t \approx 10^4\ T_{orb}(R_{in})$. The displayed results correspond to 1 realization of the models. In the case of model B1, the parameters are the same as in Fig. 1.8, except for $M_c(R_{out}) = 0.01\ M_\bullet$ (see also Table 2.1)

- the SMBH of mass $M_\bullet$ is considered to be a fixed source of static Keplerian potential, $\phi_\bullet(r) = -GM_\bullet/r$,
- the stellar disc is modelled as a group of N_d interacting stars of mass m_d; the semi-major axes of the stellar orbits in the disc are initially distributed according to $dN/da \propto a^{\alpha_d}$, $a \in \langle a_{d,min}, a_{d,max} \rangle$; the disc is considered to be initially thin with half-opening angle Δ,

- the spherical star cluster is treated as N_c interacting stars of mass m_c; initial distribution of the semi-major axes of the orbits in the cluster obeys $dN/da \propto a^{\alpha_c}$, $a \in \langle a_{c,min}, a_{c,max} \rangle$.

Temporal evolution of this stellar system is followed numerically, by means of the N-body integration code NBODY6 (Aarseth 2003). The Keplerian potential of the SMBH has been incorporated into the publicly available version of the code as an additional external potential. The original code has also been optimized for integrations of the massive central body dominated stellar systems in cooperation with its author, Sverre J. Aarseth. The xy-plane of our Cartesian reference frame is defined by the plane of symmetry of the disc. The stellar motions in the disc are considered to be initially prograde, i.e. the reference z-axis is pointing to the same hemisphere as the initial mean angular momentum of the disc.

We start with a particular setup (hereafter 'canonical' configuration; see also Table 2.1) that is primarily motivated by the stellar system observed in the Galactic Centre and discuss the acquired results later on (see Sect. 2.4). For simplicity, the stellar orbits in both the cluster and the disc are constructed to be initially geometrically circular. Although this setup is not very realistic for the orbits in the cluster, it can still provide useful insights into the evolution of general non-circular stellar systems.

The parameters of the disc are the same as in the previous section. In particular, the individual stars are assumed to have one of the following three masses: $m_{d1} = 2.5 \times 10^{-6}\,M_\bullet$, $m_{d2} = 7.5 \times 10^{-6}\,M_\bullet$, $m_{d3} = 2.5 \times 10^{-5}\,M_\bullet$. The corresponding abundances are $N_{d1} = 600$, $N_{d2} = 300$ and $N_{d3} = 100$, yielding the total mass of the disc $N_{d1}m_{d1} + N_{d2}m_{d2} + N_{d3}m_{d3} = 6.25 \times 10^{-3}\,M_\bullet$. The initial distribution of the orbital elements is independent of the stellar mass. The power-law index for the initial distribution of the radii, R, of the orbits in the disc is set to $\alpha_d = -1$. This value corresponds to the surface density profile, $\Sigma(R) \propto R^{-2}$, that has been recently reported to describe the stellar disc observed in the innermost parsec of our Galaxy (Paumard et al. 2006, see also Chap. 4 of this thesis for details). The initial interval for the radii of the orbits spreads over one order of magnitude, $R \in \langle R_{in}, 10R_{in} \rangle$, where R_{in} denotes the initial inner radius of the disc. The disc is considered to have the initial half-opening angle $\Delta = 2.5°$. The initial distributions of the orbital nodal longitudes, Ω, and arguments of pericentre, ω, are uniform, in accord with the assumption that the disc is initially axially symmetric.

The cluster consists, in the canonical configuration, of $N_c = 1.25 \times 10^4$ equal-mass stars with $m_c \approx 1.9 \times 10^{-6}\,M_\bullet$. The total mass of the cluster thus is $N_c m_c \approx 0.02\,M_\bullet$. The initial distribution of the radii of the orbits in the cluster, $R \in \langle 0.5\,R_{in}, 20\,R_{in} \rangle$, obeys the power-law with index $\alpha_c = 1/4$ which corresponds to the Bahcall-Wolf radial density profile (1.30). Hence, the mass of the cluster enclosed within the initial outer radius of the disc, R_{out}, equals the characteristic mass $M_c(R_{out}) = 0.01\,M_\bullet$ considered in the previous section. Since we assume the cluster to be initially spherically symmetric, the distributions of $\cos i$, Ω and ω of the orbits in the cluster are initially uniform.

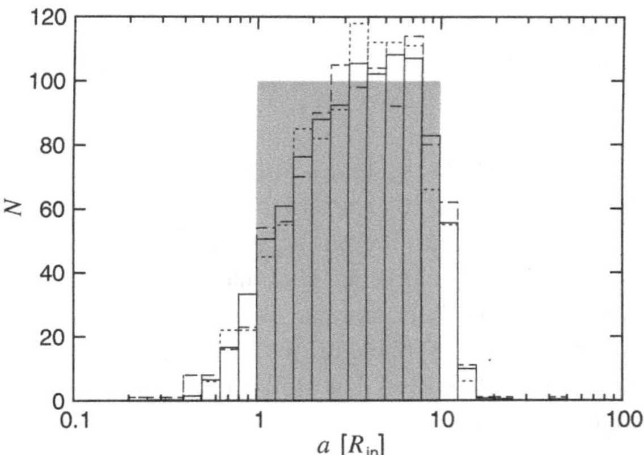

Fig. 2.3 Evolution of the orbital semi-major axes in the disc embedded in the spherical cluster in terms of their initial distribution (*grey rectangle*) and its evolved state at $t \approx 10^4\, T_{\rm orb}(R_{\rm in})$ for three different treatments of the cluster gravity: no cluster included (*dashed boxes*; see also Sect. 1.7 and Table 2.1, model B0), predefined analytic potential (*dotted boxes*; see also Sect. 2.1 and Table 2.1, model B1) and the cluster treated as a large number of gravitating stars (*solid boxes*; canonical configuration, see Table 2.1). We see that the evolved states are roughly the same in all three cases. The displayed results for the canonical configuration are averaged over 9 realizations. In the remaining two cases, the distributions describe 1 realization

2.2.1 Accelerated Exchange of Angular Momentum

Our calculations show (see also Haas and Šubr 2012) that the full N-body treatment of the cluster gravity does not affect the evolution of the distribution of the orbital semi-major axes in the disc in any noticeable way. Figure 2.3 demonstrates this finding. It shows the comparison of the initial distribution (grey rectangle) and its state at $t \approx 10^4\, T_{\rm orb}(R_{\rm in})$ in three different cases. The previously discussed results for the models with no cluster included and the cluster represented by an analytic potential are depicted by the dashed and dotted boxes, respectively (see Sects. 1.7 and 2.1 for details). The newly introduced canonical model with the full N-body treatment of the cluster gravity is denoted by the solid boxes. As we can see, the evolved distributions are similar in all three cases, which indicates that the energy transfer in the disc is dominated by its high stellar density rather than by the gravitational interaction with the embedding cluster.

On the other hand, we find that the full N-body treatment of the cluster gravity has a significant impact upon the exchange of angular momentum. Figure 2.4 shows the evolution of the root-mean-square eccentricity, $e_{\rm rms}$, (left panel) and inclination, $i_{\rm rms}$, (right panel) of the stellar orbits in the disc. We see that, although the initial phase is nearly identical in all three cases, at some moment, evolution of both elements is accelerated if the cluster is treated as a group of gravitating stars (solid lines). Our results, discussed in detail below, show that this acceleration is due to averaging over an increasing number of orbits whose eccentricity and inclination start to oscillate to

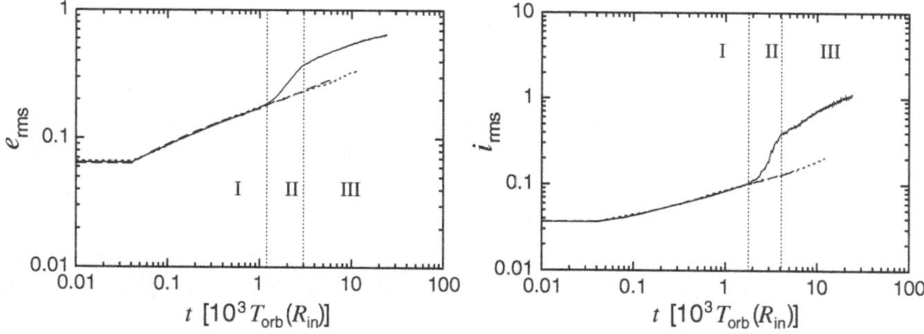

Fig. 2.4 Evolution of the root-mean-square eccentricity e_{rms} (*left panel*) and inclination i_{rms} (*right panel*) of the stellar orbits in the disc embedded in the spherical cluster for the same three treatments of the cluster gravity as in Fig. 2.3. Time is given in multiples of the orbital period that corresponds to the initial inner radius of the orbits in the disc, R_{in}. We see that in the case of the cluster modelled in the full N-body way (*solid lines*; canonical configuration), the evolution of both elements is accelerated. The displayed results for the canonical configuration are averaged over 9 realizations. In the remaining two cases, the curves describe 1 realization (model parameters are summarized in Table 2.1)

very high values. We further suggest that these oscillations are caused by the Kozai-Lidov dynamics in the combined potential of the disc and a system of flattened structures that, according to our computations, form in the cluster.

2.2.2 e-Instability in the Cluster

As we have already mentioned above, the initially spherically symmetric cluster develops macroscopic non-spherical structures. It turns out that this is due to the gravity of the embedded thin disc. In this section, we describe the formation of these structures whose gravitational potential, subsequently, influences the orbits of the stars from the disc.

In the initial phase of the evolution of the cluster (i.e. before the non-spherical structures are formed), the individual stellar orbits in the cluster can be, in principle, affected by the following processes:

- the Kozai-Lidov mechanism in the axisymmetric potential of the disc,
- negative pericentre shift due to the spherical mean potential of the cluster,
- two-body and resonant relaxation of the cluster.

In order to determine which of these processes are effective enough to have an impact on the initial phase of the cluster evolution in the canonical configuration, we evaluate their characteristic timescales for this particular setting. For this purpose, we first focus on the inner parts of the cluster where, according to our results, the formation of the macroscopic non-spherical perturbation begins. The evolution of the outer parts of the cluster is described in Sect. 2.2.6.

The characteristic Kozai-Lidov time-scale (1.13) in the case of a disc-like perturber reads $T_{KL} \sim (M_\bullet/M_d)(R_d/R)^3\, T_{orb}$, where M_d and R_d are the mass and characteristic radius of the disc, respectively, R is the radius of the orbit and $T_{orb} \sim \sqrt{R^3/GM_\bullet}$. For the canonical radial density profile of the disc, we estimate $R_d \sim R_{in}$. Since $R \sim R_{in}$ in the inner parts of the cluster, the ratio R_d/R is of order unity. In the canonical configuration, the mass of the disc, $M_d = N_{d1}m_{d1} + N_{d2}m_{d2} + N_{d3}m_{d3}$, is of order $\sim 10^{-3}\, M_\bullet$ and, therefore, $T_{KL} \sim 10^3\, T_{orb}$. The timescale for the pericentre shift (1.22) can be written as $T_c \sim (M_\bullet/N_c(R)m_c)\; T_{orb}$, where $N_c(R)$ is the number of the stars in the cluster enclosed within radius R. For $R \sim R_{in}$, we have $N_c(R_{in}) \sim 10^2$. Hence, considering the mass of the stars in the cluster $m_c \sim 10^{-6}\, M_\bullet$, we obtain $T_c \sim 10^4\, T_{orb}$. The characteristic timescale for the vector resonant relaxation in the cluster (1.33) can be, in the canonical configuration, expressed as $T_{vr,sf} \sim (M_\bullet/N_c(R)^{1/2}\, m_c)\; T_{orb}$. For $R \sim R_{in}$, we have $N_c(R_{in})^{1/2} \sim 10$ and $T_{vr,sf} \sim 10^5\, T_{orb}$. The characteristic timescales of the scalar resonant relaxation and the two-body relaxation of the cluster are even longer (see Sect. 1.6).

The initial phase of the orbital evolution of the cluster covers a period of time of order $\sim 10^3\, T_{orb}(R_{in})$, as can be inferred, e.g. from Fig. 2.6 which we describe below. Hence, neither two-body nor resonant relaxation of the cluster can significantly affect the individual stellar orbits from its inner parts during this phase of the evolution. On the other hand, the Kozai-Lidov mechanism in the axisymmetric potential of the disc is efficient enough to have an impact on these orbits. Furthermore, since $T_c \sim 10^4\, T_{orb}(R_{in})$, we cannot safely exclude the effect of the pericentre shift in the mean potential of the cluster. In fact, our calculations show that for most of the orbits in the cluster with the initial radius $> R_{in}$, the Kozai-Lidov oscillations are damped by the mean spherical potential of the cluster during the initial phase of their evolution. This is demonstrated in Fig. 2.5 which displays the evolution of the orbital elements for a typical star from the innermost parts of the cluster (initial radius of the orbit $\approx R_{in}$). As we can see, both eccentricity e (dashed line) and inclination i (solid line) remain roughly constant until $t \approx 2 \times 10^3\, T_{orb}(R_{in})$. The nodal longitude Ω (dotted line) and argument of pericentre ω (dot-dashed line) of the orbit show a secular rotation. In the case of the nodal longitude, the rotation is, for this particular star, rather slow due to the high inclination of the stellar orbit since $d\Omega/dt \propto \cos i \to 0$ for $i \to \pi/2$ (see Eqs. (1.12)). In conclusion, we find that, in the initial phase of the evolution of the cluster, the stellar orbits from its inner parts undergo combined precession of their nodal and apsidal lines only, keeping their eccentricity and inclination close to their initial values.

Our results further show that the orbits from the inner parts of the cluster tend to change their orientations in a way such that their eccentricity vectors, $\boldsymbol{e}$, are pointing in similar directions, parallel to the plane of the disc. Figure 2.6 illustrates this finding by means of sinusoidal projection of eccentricity vectors of the stars from the cluster whose osculating semi-major axis, a, fulfils $a < 1.5\, R_{in}$. In the initial state (top-left panel), the orbits are distributed uniformly over the whole plot, which is in accord with the assumed initial spherical symmetry of the cluster. In the other panels, however, we observe that most of the orbits tend to form a rather compact

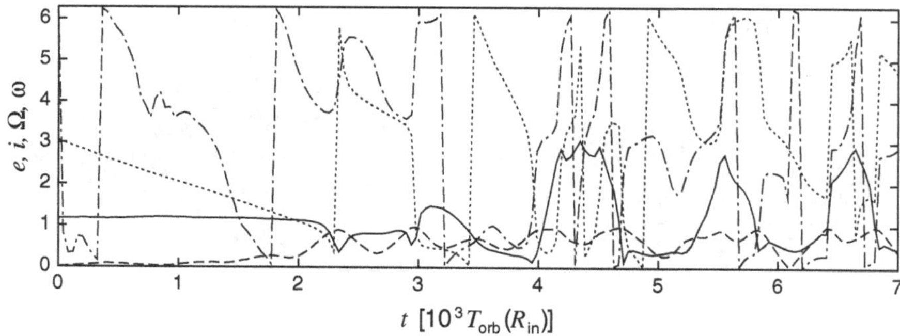

Fig. 2.5 Evolution of the orbital eccentricity e (*dashed line*), inclination i (*solid line*), nodal longitude Ω (*dotted line*) and argument of pericentre ω (*dot-dashed line*) of a typical stellar orbit from the innermost parts of the cluster in the canonical configuration (parameters summarized in Table 2.1). We see that, in the initial phase of the evolution, the Kozai-Lidov oscillations are damped by the mean spherical potential of the cluster ($t \lesssim 2 \times 10^3\ T_{\rm orb}(R_{\rm in})$). When the $\boldsymbol{e}$-structure in the cluster is formed, the orbit undergoes its Kozai-Lidov cycles ($2 \times 10^3\ T_{\rm orb}(R_{\rm in}) \lesssim t \lesssim 3.5 \times 10^3\ T_{\rm orb}(R_{\rm in})$). Subsequently, it is captured by the $\boldsymbol{e}$-structure in the disc, thus showing typical oscillations of eccentricity and inclination ($t \gtrsim 3.5 \times 10^3\ T_{\rm orb}(R_{\rm in})$)

group which is located on the equatorial line ($i_{\boldsymbol{e}} = 90°$) of the plots. Hereafter, we refer to this type of orbital structure, as the $\boldsymbol{e}$-structure. We point out that, despite the similar orientations of the eccentricity vectors of the individual orbits in the $\boldsymbol{e}$-structure, the initial spherical symmetry of the mass distribution in the cluster is disturbed only slightly. Nevertheless, our results indicate that the small deviation from the initial symmetry is sufficient for the $\boldsymbol{e}$-structure to have a significant impact upon the evolution of the disc. Let us also note that the two horizontal features visible in the top-right and bottom-left panels of Fig. 2.6 are only temporary and we explain their origin in Sect. 2.3.1.

Although the process of formation of the $\boldsymbol{e}$-structure is not fully clear yet, we suggest that it can be understood in the following way. As mentioned above, the orbits from the cluster undergo combined nodal and apsidal precession. Since the rate of this precession depends on the orbital elements, individual orbits precess at different rates, which leads to a change of their relative orientations. Consequently, also the strength of the long-term mutual interaction of the stars on these orbits is changed. We suggest that, once two (or more) orbits achieve a specific orientation in which their mutual interaction is strong enough, they tend to dynamically couple together and precess further synchronously. Furthermore, once the orbits become dynamically coupled, the probability of capture of another orbit increases as the potential well of multiple coupled orbits is deeper than those of the individual orbits. As a result, coupling of the individual orbits turns into $\boldsymbol{e}$-instability and the initially spherically symmetric cluster gradually forms a single $\boldsymbol{e}$-structure that includes most of the stars from the affected region.

According to our calculations, the $\boldsymbol{e}$-structure slowly rotates around the symmetry axis of the disc on a timescale of the order of magnitude $\sim 10\ T_{\rm KL}(R_{\rm in})$. Although the

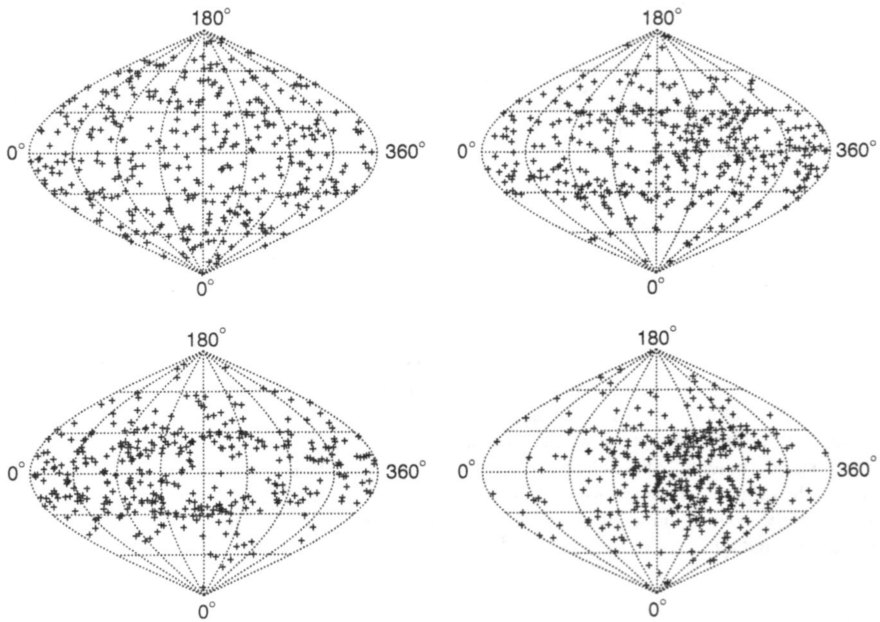

Fig. 2.6 Directions of the eccentricity vectors of the individual stars from the cluster with the osculating semi-major axes $a < 1.5\,R_{\rm in}$ in terms of angles Ω_e (abscissa) and i_e (ordinate) in sinusoidal projection. The plots describe the state at four different times: $t = 0$ (*top-left panel*), $t \approx 0.6 \times 10^3\,T_{\rm orb}(R_{\rm in})$ (*top-right panel*), $t \approx 1.2 \times 10^3\,T_{\rm orb}(R_{\rm in})$ (*bottom-left panel*), and $t \approx 1.8 \times 10^3\,T_{\rm orb}(R_{\rm in})$ (*bottom-right panel*). The initial state corresponds to the initial spherical symmetry of the cluster. Subsequently, we observe formation of the $\boldsymbol{e}$-structure (the compact group in the plots). The two horizontal features in the in the *top-right* and *bottom-left panel* are temporary (see Sect. 2.3.1). Model parameters are set to their canonical values (see Table 2.1)

rate of rotation of the individual orbits strongly fluctuates around the mean rotation of the $\boldsymbol{e}$-structure as a whole, these fluctuations are not strong enough to disrupt the ongoing formation of the $\boldsymbol{e}$-structure. The direction in which the $\boldsymbol{e}$-structure rotates depends on the mean inclination, $\langle i \rangle$, of the orbits in the $\boldsymbol{e}$-structure. In particular, we find that the rotation is opposite for $\langle i \rangle < \pi/2$ and $\langle i \rangle > \pi/2$. Furthermore, our results show that the eccentricities and inclinations of the orbits near the outer edge of the $\boldsymbol{e}$-structure are similar to the initial conditions for the cluster, i.e. these orbits are rather low-eccentric and the distribution of their $\cos i$ is roughly uniform, yielding the mean inclination of $\approx\pi/2$ (see the crosses in the right panel of Fig. 2.7). On the other hand, the orbits with smaller semi-major axes (circles in the right panel of Fig. 2.7) have higher eccentricities. We suggest that this can be explained by the following argument.

As we have already mentioned, the spherically symmetric mean potential of the cluster suppresses, until the $\boldsymbol{e}$-structure is formed, the Kozai-Lidov mechanism in the potential of the disc. Hence, the individual orbits from the cluster retain their eccentricity and inclination close to their initial values. Upon the formation of the

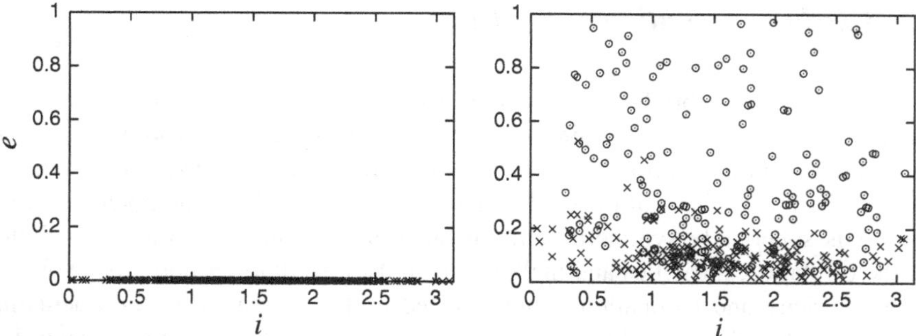

Fig. 2.7 Eccentricity e and inclination i of the orbits from the innermost parts of the cluster: orbits with the osculating semi-major axis $a < R_{\rm in}$ and $R_{\rm in} < a < 1.5\,R_{\rm in}$ are denoted by *circles* and *crosses*, respectively. The initial state is shown in the *left panel*. The *right panel* describes the state at $t \approx 1.2 \times 10^3\,T_{\rm orb}(R_{\rm in})$. We see that the orbits with larger osculating semi-major axes (*crosses*) are still low-eccentric even when a significant $\boldsymbol{e}$-structure is formed (cf. the *bottom-left panel* of Fig. 2.6). On the other hand, the orbits with smaller semi-major axes are already oscillating (*circles*). Model parameters are set to their canonical values (see Table 2.1)

$\boldsymbol{e}$-structure, however, the spherical symmetry of the mean potential is disturbed and, therefore, its damping effect is weaker. Consequently, the Kozai-Lidov mechanism in the potential of the disc starts to affect the orbits in the $\boldsymbol{e}$-structure which thus undergo the high-amplitude Kozai-Lidov oscillations of eccentricity and inclination. As the Kozai-Lidov timescale (1.13) depends strongly upon the semi-major axis and, furthermore, since the $\boldsymbol{e}$-instability gradually propagates from the centre of the cluster outwards (see Sect. 2.2.6 for details), the orbits in the central parts of the $\boldsymbol{e}$-structure are already oscillating while the elements of the orbits near its outer edge are still unaffected.

The onset of the Kozai-Lidov mechanism in the potential of the disc after the formation of the $\boldsymbol{e}$-structure is also visible in Fig. 2.5. In particular, we see that all of the displayed orbital elements evolve, in the interval $2 \times 10^3\,T_{\rm orb}(R_{\rm in}) \lesssim t \lesssim 3.5 \times 10^3\,T_{\rm orb}(R_{\rm in})$, in a way that corresponds to the lower librational lobe (around $\omega = 3\pi/2$) in the Kozai-Lidov diagrams (see Sect. 1.3, Figs. 1.2 and 1.3).

In the context of these findings, we argue that the $\boldsymbol{e}$-structure represents, shortly after its formation, a flattened stellar overdensity which is roughly perpendicular to the plane of the disc. Moreover, due to the fact that the frequencies of the nodal and apsidal precessions in the Kozai-Lidov mechanism are similar, the orbits from the $\boldsymbol{e}$-structure that subsequently undergo their Kozai-Lidov cycles roughly preserve the orientation of their apsidal lines (the small deviations are 'absorbed' by the other orbits due to their mutual interaction). Consequently, the coherent rotation of the $\boldsymbol{e}$-structure is not disturbed and, therefore, the Kozai-Lidov oscillations of eccentricity and inclination of the orbits in the $\boldsymbol{e}$-structure do not disturb its overall shape. The orbits are either highly inclined and low-eccentric or they are found in the plane of the disc but have very high eccentricities. As a result, the slowly rotating $\boldsymbol{e}$-structure remains flattened and its potential causes, in return, the Kozai-Lidov mechanism in the disc.

2.2.3 Orbital Evolution of the Disc

Now, that we have briefly outlined the orbital evolution of the cluster, we have sufficient information to explain the accelerated increase of the root-mean-square eccentricity and inclination of the stellar orbits from the disc observed in Fig. 2.4. During the initial phase of the evolution (denoted as 'I'), the mean potential of the cluster is spherically symmetric and, as such, has no impact upon the evolution of the orbital eccentricities and inclinations in the disc. Furthermore, the evolution of eccentricity and inclination is not affected by the relaxation in the fluctuating potential of the cluster as this process operates on timescales that are by two orders of magnitude longer than the initial phase of the evolution itself. Hence, during this phase, the disc evolves solely due to its two-body relaxation whose characteristic timescale (1.36) is very short since the orbits in the disc are initially nearly circular. The evolution of the root-mean-square eccentricity and inclination of the orbits from the disc is thus similar to the evolution found in the cases with the cluster emulated by the predefined analytic potential and with no cluster present at all.

When the $\boldsymbol{e}$-structure in the cluster is formed, however, its potential starts to induce the Kozai-Lidov oscillations with combined nodal and apsidal precession of the orbits in the disc with respect to the plane of the $\boldsymbol{e}$-structure. Since this plane is roughly perpendicular to the plane of the disc with respect to which we evaluate the orbital elements, the nodal precession with respect to the plane of the $\boldsymbol{e}$-structure transforms to extreme oscillations of inclination over the whole interval $\langle 0, \pi \rangle$ with respect to the plane of the disc (see Sect. 2.3.3 for details). Averaging of eccentricity and inclination over an increasing number of such oscillating orbits then leads to the accelerated increase of their root-mean-square values observed in Fig. 2.4 (phases 'II' and 'III'). The rate of this increase is determined by the radial density profile of the disc.

The existence of two different phases of the accelerated evolution of the root-mean-square eccentricity and inclination of the orbits in the disc can be understood from Fig. 2.8 which shows the evolution of eccentricity e and inclination i of two orbits from the innermost parts of the disc. We see that, for both orbits, the oscillations of inclination do not immediately cover the whole interval $\langle 0; \pi \rangle$ but only a small fraction of it (hereafter 'basic' mode). Furthermore, the full amplitude of the oscillations of inclination (hereafter 'extreme' mode) is reached at different times for each of the two orbits ($t \approx 5 \times 10^3\, T_{\rm orb}(R_{\rm in})$ and $t \approx 3 \times 10^3\, T_{\rm orb}(R_{\rm in})$ in the top and bottom panel, respectively). Eccentricity of the orbits oscillates to very high values already in the basic mode. However, in the extreme mode, the orbits spend notably longer periods of time in their high-eccentricity states. This behaviour is typical for most of the orbits from the affected parts of the disc. We thus suggest that the first phase of the accelerated evolution of the root-mean-square eccentricity and inclination observed in Fig. 2.4 (phase 'II') corresponds to the averaging over the oscillations in the basic mode. When most of the orbits in the affected parts of the disc are already oscillating in the basic mode, the root-mean-square eccentricity and inclination saturate, further showing slower increase due to averaging over an increasing number of orbits that start to oscillate in the extreme mode (phase 'III').

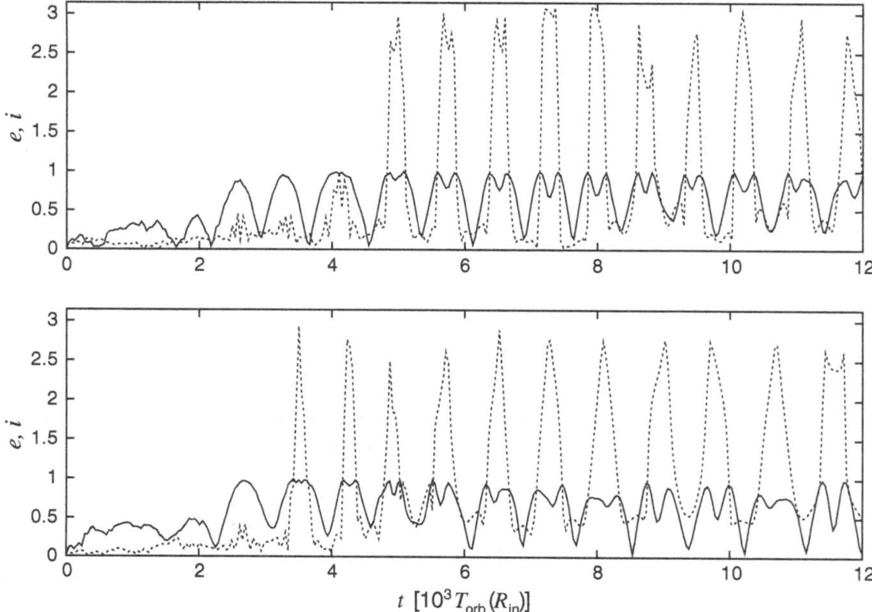

Fig. 2.8 Evolution of the orbital eccentricity e (*solid lines*) and inclination i (*dashed lines*) of two typical stars from the innermost parts of the disc within the canonical configuration (see Table 2.1). We observe that both elements undergo extreme oscillations. The maximum value of eccentricity, $e \to 1$, is reached when the orbits are roughly perpendicular to the parent disc: $i \approx \pi/2$. We further see that both orbits start their oscillations (basic mode) at approximately the same time: $t \approx 2 \times 10^3\, T_{\rm orb}(R_{\rm in})$. It also turns out that, for each of the orbits, the oscillations of inclination reach the extreme amplitude $\langle 0; \pi \rangle$ at different times

2.2.4 *e*-Instability in the Disc

Similarly, to the case of the cluster in the axisymmetric potential of the disc, our calculations reveal that also the disc under the effect of the flattened potential of the **e**-structure in the cluster evolves its own **e**-structure. This finding is demonstrated in Fig. 2.9 which shows the distribution of directions of the eccentricity vectors of the innermost orbits in the disc in terms of angles Ω_e and i_e at $t = 0$ (left panel) and $t \approx 6 \times 10^3\, T_{\rm orb}(R_{\rm in})$ (right panel). We observe that, in accord with the assumed initial axial symmetry of the disc, the distribution in the left panel is uniform with respect to Ω_e (abscissa). Furthermore, since the disc is assumed to be initially thin, the individual vectors are confined to a narrow belt along the equatorial line of the plot. In the right panel, however, we can see that nearly all of the displayed orbits belong to a single compact group and, therefore, they form a significant **e**-structure. Despite the similar orientation of the apsidal lines of the orbits in the **e**-structure, the initial overall axial symmetry of the mass distribution in the disc does not appear

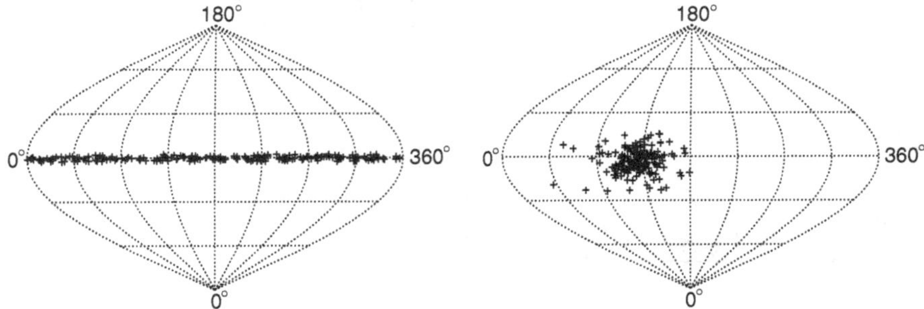

Fig. 2.9 Directions of the eccentricity vectors of the orbits from the disc with the osculating semi-major axis $a < 1.5\,R_{\rm in}$ in terms of angles Ω_e (abscissa) and i_e (ordinate) in sinusoidal projection. The initial state is displayed in the *left panel* while the *right panel* describes the directions at $t \approx 6 \times 10^3\,T_{\rm orb}(R_{\rm in})$. We see that, in the evolved state, nearly all of the displayed orbits belong to a single compact group, thus forming a significant $\boldsymbol{e}$-structure. Model parameters are set to their canonical values (see Table 2.1)

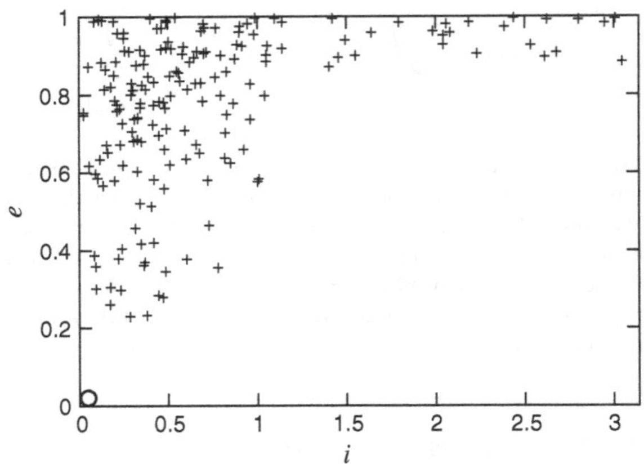

Fig. 2.10 Eccentricity e and inclination i of the innermost orbits in the evolved disc ($t \approx 6 \times 10^3\,T_{\rm orb}(R_{\rm in})$; only orbits with the osculating semi-major axis $a < 1.5\,R_{\rm in}$ are displayed). We see that the orbits are mostly low-inclined and highly eccentric. The initial state is denoted by the *empty circle*. Model parameters are set to their canonical values (see Table 2.1)

to be disturbed dramatically by the $\boldsymbol{e}$-structure formation (see the panels in the left column of Fig. 2.11), similar to the case of the $\boldsymbol{e}$-structure in the cluster.

Our results further show that this $\boldsymbol{e}$-structure is, unlike the $\boldsymbol{e}$-structure in the cluster, formed by rather eccentric and mostly low-inclined orbits (see Fig. 2.10). On the other hand, similar to the $\boldsymbol{e}$-structure in the cluster, the $\boldsymbol{e}$-structure in the disc also rotates around the initial symmetry axis of the disc. Furthermore, we find that the individual orbits from the $\boldsymbol{e}$-structure in the disc rotate much more synchronously in comparison with the orbits from the $\boldsymbol{e}$-structure in the cluster. In other words, the rotation of the $\boldsymbol{e}$-structure in the disc much more resembles the rotation of a solid body (see Fig. 2.11). We attribute this finding to the fact that the stellar density in the initially thin disc is higher than in the spherical cluster. Consequently, the $\boldsymbol{e}$-structure in the disc represents a deeper potential well that bounds the individual orbits more

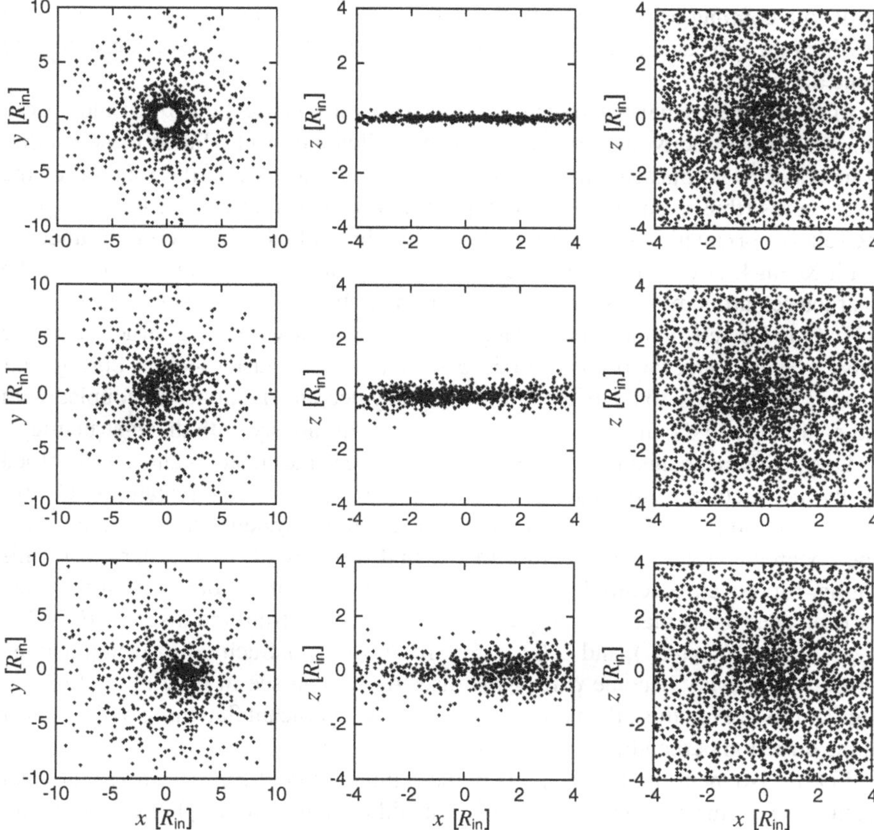

Fig. 2.11 The positions of the stars in the disc (*left* and *middle column*) and the cluster (*right column*) in projection onto the xy-plane (*left column*) and the xz-plane (*middle* and *right column*; zoomed on the central parts of the disc and the cluster) at three different times: $t = 0$ (*top panels*), $t \approx 2.8 \times 10^3\, T_{\rm orb}(R_{\rm in})$ (*middle panels*; corresponds to the time when the $\boldsymbol{e}$-structures in the cluster and the disc reach similar orientation) and $t \approx 10.9 \times 10^3\, T_{\rm orb}(R_{\rm in})$ (*bottom panels*). Model parameters are set to their canonical values (see Table 2.1)

strongly than in the case of the $\boldsymbol{e}$-structure in the cluster. This appears to be crucial for the mutual interaction of these two $\boldsymbol{e}$-structures.

2.2.5 Interaction of the e-Structures in the Cluster and the Disc

We have learned that both the cluster and the disc form, during their orbital evolution, significant $\boldsymbol{e}$-structures slowly rotate around the symmetry axis of the disc. It further turns out that the rate of this rotation is different for the two $\boldsymbol{e}$-structures, which leads to a continuous change of their relative orientation and, eventually, to a state in which the typical eccentricity vectors in the two ***e***-structures point in roughly the same direction.

Furthermore, we have also seen that the ***e***-structures cause the Kozai-Lidov oscillations of eccentricity and inclination of the individual orbits. Since these oscillations occur on a timescale that is shorter than the timescale of the rotation of the ***e***-structures (thus even shorter in comparison with the timescale of the change of the relative orientation of the ***e***-structures), the individual orbits certainly undergo a full Kozai-Lidov cycle during the period of time in which the orientations of the ***e***-structures are similar. As we have mentioned above, the ***e***-structure in the disc appears to be self-bound more strongly in comparison with the ***e***-structure in the cluster. Hence, once the orbits from the ***e***-structure in the cluster incline, during their Kozai-Lidov cycles, to the plane of the disc, they become dynamically coupled with the ***e***-structure in the disc, precessing further synchronously with this ***e***-structure.

Due to this mechanism, the innermost ***e***-structure in the cluster is entirely absorbed by the ***e***-structure in the disc as soon as they reach similar orientations. Consequently, the innermost parts of the investigated stellar system further contain only one coherently rotating ***e***-structure that includes nearly all of the stars from this region. Moreover, eccentricity and inclination of some of the orbits from the cluster even start to oscillate in a way typical for the orbits from the disc (cf. Figs. 2.5, $t \gtrsim 3.5 \times 10^3\, T_{\rm orb}(R_{\rm in})$, and 2.8, $t \gtrsim 5 \times 10^3\, T_{\rm orb}(R_{\rm in})$). Such orbits from the cluster thus effectively become part of the disc. As we can see in the right column of Fig. 2.11, however, this effect does not lead to any significant flattening of the cluster in the direction perpendicular to the plane of the disc.

Let us also mention that the plots in the right column of Fig. 2.11 reveal another feature of the cluster evolution. In particular, although the initial spherical symmetry of the core of the cluster is disturbed only slightly during the evolution, in the evolved state, the core is not centred precisely on the SMBH anymore.

2.2.6 Propagation of the e-Instability to Farther Regions

So far, we have been investigating the evolution of the inner parts of the stellar system. However, our calculations show that the ***e***-instability in the cluster propagates also to its outer parts. This effect is demonstrated in Fig. 2.12 which shows the eccentricity vector angle Ω_e of the stars in the cluster, depending on their instantaneous distance from the central SMBH at four different stages of the orbital evolution of the cluster. As we can see in the top-left panel, the initial distribution is in accord with the assumed initial spherical symmetry of the cluster. On the other hand, the remaining three panels show overdensities which propagate outwards in the plot and indicate presence of ***e***-structures even in the outer parts of the cluster.

Furthermore, similar to the innermost region of the cluster, the flattened potential of the disc leads to the Kozai-Lidov oscillations of eccentricity and inclination of the individual orbits in the more distant regions of the cluster as soon as the spherical

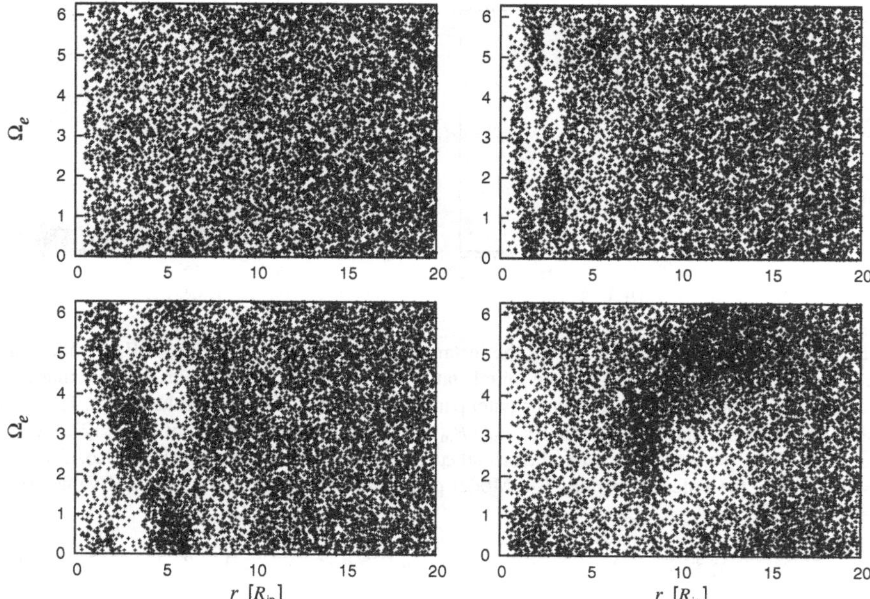

Fig. 2.12 Values of the eccentricity vector angle Ω_e and the instantaneous distance r from the SMBH of the stars in the cluster at four different stages of the cluster evolution: $t = 0$ (*top-left panel*), $t \approx 1.2 \times 10^3\ T_{\rm orb}(R_{\rm in})$ (*top-right panel*), $t \approx 2.4 \times 10^3\ T_{\rm orb}(R_{\rm in})$ (*bottom-left panel*) and $t \approx 12 \times 10^3\ T_{\rm orb}(R_{\rm in})$ (*bottom-right panel*). We see that the $\boldsymbol{e}$-instability gradually propagates to the outer parts of the cluster. Model parameters are set to their canonical values (see Table 2.1)

symmetry of the mean potential is disturbed by the formation of the $\boldsymbol{e}$-structures. As the potential of the disc in these regions is weaker, however, the corresponding timescale of the oscillations is longer and, therefore, the eccentricity of the orbits in the outermost $\boldsymbol{e}$-structures is still quite low (see Fig. 2.13).

Finally, let us mention that, unlike the cluster, the disc forms only one dominating $\boldsymbol{e}$-structure that includes nearly all of its stars, as can be inferred from Fig. 2.14.

2.3 Basic Processes

In the previous section, we have described the complex evolution of the stellar system that consists of an initially thin disc which is embedded in an extended spherical cluster, both centred on the SMBH, by means of the full N-body modelling. In the following, we investigate the individual basic processes included in this evolution separately, by means of simplified models.

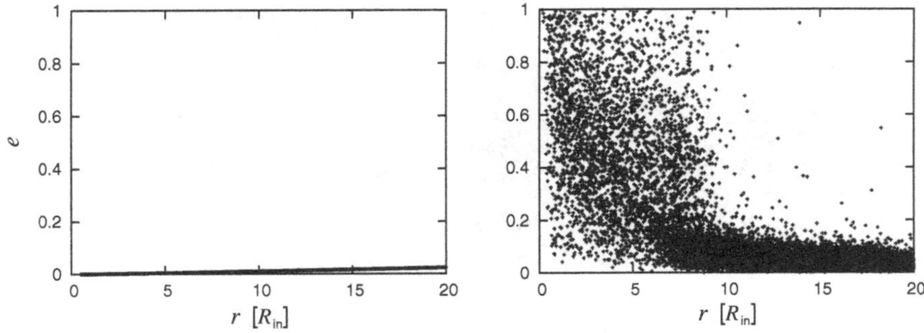

Fig. 2.13 Values of eccentricity e and the instantaneous distance r from the SMBH of the stars in the cluster. The left panel shows the initial state when the orbits are geometrically circular (the Keplerian eccentricity of the orbits in the outer parts of the cluster is thus larger than zero). The right panel describes the state at $t \approx 12 \times 10^3\ T_{\rm orb}(R_{\rm in})$ (cf. the bottom-right panel of Fig. 2.12). We see that, in the inner parts of the cluster, the orbital eccentricities are increased due to the Kozai-Lidov oscillations in the potential of the disc. Model parameters are set to their canonical values (see Table 2.1)

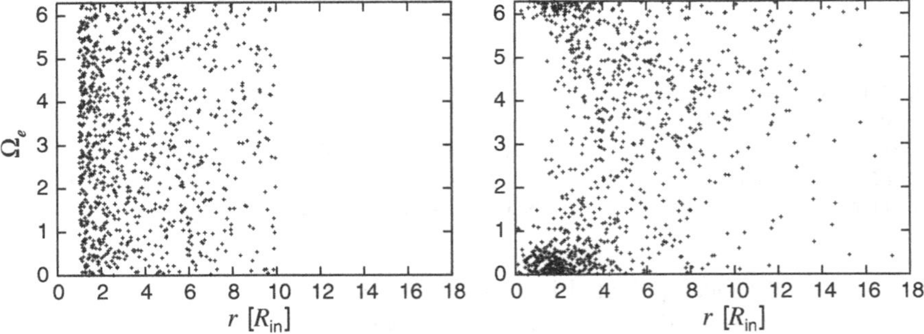

Fig. 2.14 Values of the eccentricity vector angle Ω_e and the instantaneous distance r from the SMBH of all the stars in the disc at $t = 0$ (*left panel*) and $t \approx 12 \times 10^3\ T_{\rm orb}(R_{\rm in})$ (*right panel*). We observe that nearly all of the stars in the disc are concentrated in one large **e**-structure. Model parameters are set to their canonical values (see Table 2.1)

2.3.1 Kozai-Lidov Dynamics in the Cluster

We first analyse the orbital evolution of an initially spherically symmetric cluster of test particles whose motions around the dominating central SMBH are perturbed by a predefined distant infinitesimally thin ring (hereafter K0a-model; see Table 2.1). In other words, we study the statistical effects of the Kozai-Lidov dynamics of the individual stars from the cluster in the potential of the ring. For this purpose, we follow the orbital evolution of the cluster numerically, by means of the N-body integration code Mbody (Šubr 2006) which is more suitable for this particular setting. The use of an independent numerical integrator also enables us to test our results for systematic flaws that might have been introduced by specific numerical methods implemented in the NBODY6 code used in the case of the canonical configuration. The effect of test particles is reached by decreasing the mass of the individual stars from the cluster to an extremely low value, $m_c \sim 10^{-12}\, M_\bullet$.

Depending on the initial conditions of the individual orbits from the cluster, the corresponding Kozai-Lidov diagrams may be of two qualitatively different topologies (see Fig. 1.2): (i) solely rotational diagrams, or (ii) diagrams with the librational and the outer rotational region. In the later case, the diagrams contain the separatrix contour which resembles a straight radial line near the point of its intersection at $e = 0$. Along the isocontours in the vicinity of these parts of the separatrix, the orbital argument of pericentre thus remains roughly constant (in both the librational and the outer rotational region). As the cluster is assumed to be initially spherically symmetric, the distribution of $\cos i$ of the stellar orbits is uniform. Furthermore, within the K0a-model, we consider the orbits to be initially near-circular and, therefore, also the initial distribution of the Kozai integral of motion (1.11) in the cluster is uniform. Since the value of the Kozai integral that corresponds to the Kozai limit is $c_1 = \sqrt{3/5} \approx 0.8$, we see that the Kozai-Lidov diagrams for most of the orbits in the cluster contain both the librational and outer rotational region. As a result, the Kozai-Lidov cycles for the majority of the orbits in the cluster include the phase with nearly constant argument of pericentre. This indicates that, during the evolution of the cluster of test particles, an overabundance of orbits with a specific value of ω might appear.

Given the diagrams with both the librational and the outer rotational regions, the Kozai-Lidov Eq. (1.32) imply that a stellar orbit with initial argument of pericentre $\omega_0 \approx 0$ or π has $(\mathrm{d}\omega/\mathrm{d}t)_0 > 0$. We thus understand that the evolutionary tracks of all the orbits in the outer rotational region follow the isocontours of the diagram in the counter-clockwise direction. Similarly, as we can deduce for initial $\omega_0 \approx \pi/2$ or $3\pi/2$, the same direction is followed for all the orbits that belong to either of the librational lobes. Since the initially near-circular orbits form a compact group in the centre of the diagram, the overabundant values of ω should, in the first phase of the evolution, correspond to the straight part of the separatrix in the first and the third quadrant, i.e. $\omega_{\text{const}} \approx 1$ and ≈ 4, respectively (see the middle panels of Fig. 2.15). Subsequently, as the orbits in the cluster shift in the diagram due to their evolution, the surroundings of the other two branches of the separatrix in the second and the fourth quadrant of the diagram become populated as well and, therefore, another two overabundances at $\omega_{\text{const}} \approx 2$ and ≈ 5 appear (bottom panels of Fig. 2.15).

Figure 2.16 demonstrates the non-uniform distribution of the orbital arguments of pericentre in the cluster in terms of the eccentricity vector angles Ω_e (abscissa) and i_e (ordinate). In particular, the right panel displays the state that corresponds to the bottom panels of Fig. 2.15, i.e. the state in which all four overabundances are already evolved. For geometrical reasons, the orbits with $\omega \approx 1$ and ≈ 2 have the same value of $i_e \approx 60°$, forming the lower horizontal belt in the right panel of Fig. 2.16. The orbits with $\omega \approx 4$ and ≈ 5 then form the upper belt.

Given the same initial conditions for the cluster, the results would be somewhat different if the distant thin ring were replaced by a disc-like perturbation. As we have mentioned in Sect. 1.3, in such a case, the Kozai-Lidov diagrams may, for

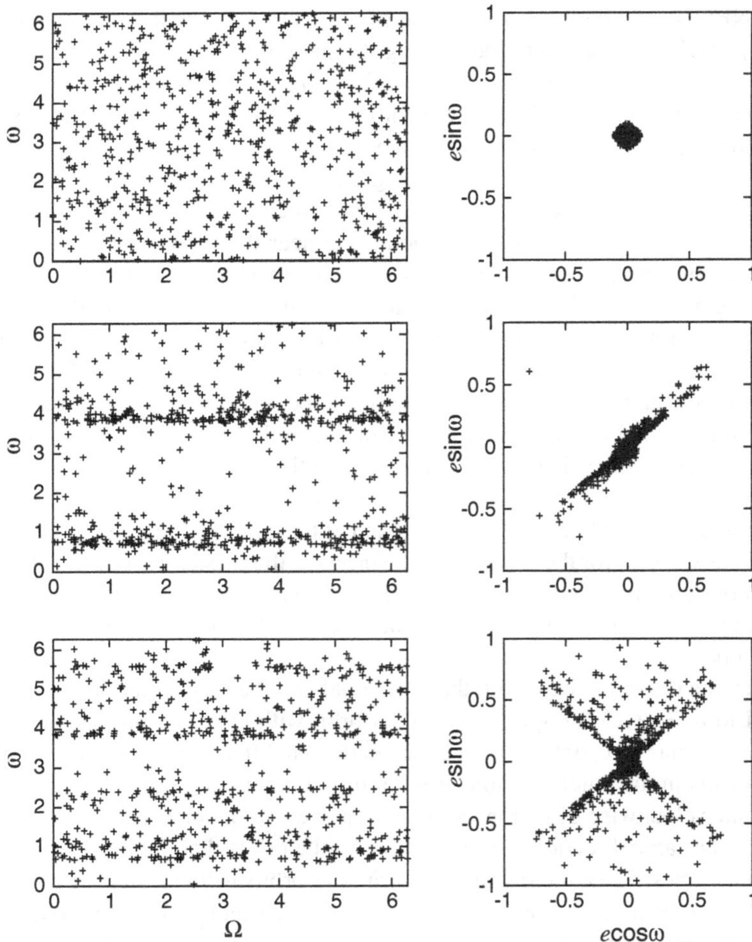

Fig. 2.15 Formation of the overabundances of specific values of the orbital argument of pericentre ω in the cluster of test particles under the perturbative gravitational influence of the distant thin ring (K0a-model; see Table 2.1). The *left panels* demonstrate the evolution of the distribution of argument of pericentre ω and longitude of the ascending node Ω of all the orbits in the cluster. The *right panels* show the position of the orbits in the Kozai-Lidov diagrams. In the initial state (*top panels*), the distribution of both ω and Ω is uniform and all the orbits are found near the origin, $e = 0$, of the diagram. In the evolved state displayed in the *middle panels* ($t \approx 2.4 \times 10^3\ T_{\rm orb}(R_{\rm in})$), we see the overabundances of $\omega \approx 1$ and ≈ 4 which correspond to the branches of the separatrix contour in the first and the third quadrant of the diagram, respectively. At a later stage of the evolution ($t \approx 5.6 \times 10^3\ T_{\rm orb}(R_{\rm in})$; *bottom panels*), also the overabundances of $\omega \approx 2$ and ≈ 5 are visible. The distribution of the nodal longitude Ω is uniform in all states, which is in accord with the axial symmetry of the problem

low enough values of the Kozai integral (1.11), contain the inner rotational region around the origin $\boldsymbol{e} = 0$. Hence, the initially near-circular orbits that have such

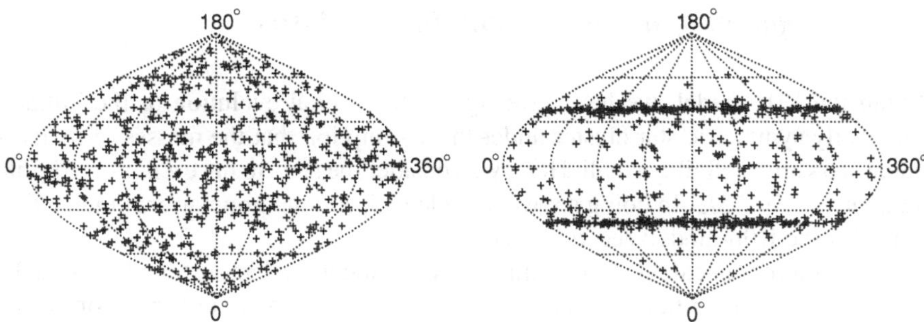

Fig. 2.16 Directions of the eccentricity vectors of all the orbits in the cluster of test particles in terms of angles Ω_e (abscissa) and i_e (ordinate) in sinusoidal projection. The initial distribution (*left panel*) is in accord with the assumed initial spherical symmetry of the cluster. In the evolved state ($t \approx 5.6 \times 10^3\, T_{\rm orb}(R_{\rm in})$; *right panel*), we see two horizontal belts (cf. the *bottom panels* of Fig. 2.15 and the *right panel* of Fig. 2.6). The model parameters are set to values that correspond to the K0a-model (see Table 2.1)

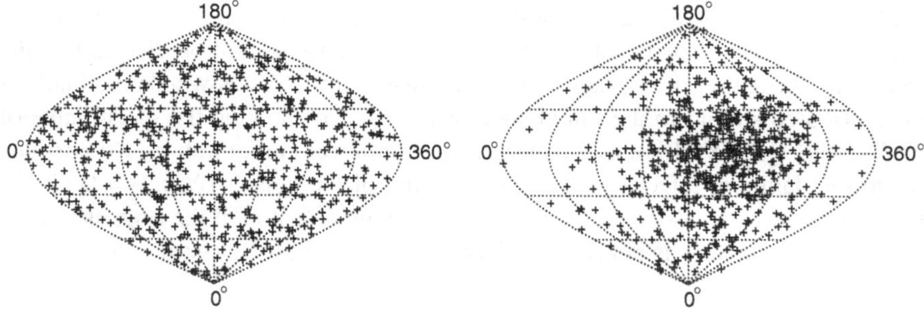

Fig. 2.17 Directions of the eccentricity vectors of all the orbits in the self-gravitating cluster in terms of angles Ω_e (abscissa) and i_e (ordinate) in sinusoidal projection. The initial distribution (*left panel*) is in accord with the assumed initial spherical symmetry of the cluster. In the evolved state ($t \approx 2.4 \times 10^3\, T_{\rm orb}(R_{\rm in})$; *right panel*), nearly all of the orbits belong to a single ***e***-structure. Model parameters correspond to the K0b-model (see Table 2.1)

values of c cannot reach the critical location in the diagram as they are confined to the inner rotational region. Such orbits thus do not contribute to formation of the overabundances. Consequently, the significance of these features is lower as they are generated by a lower number of orbits with moderate values of c.

Overabundances similar to those described above have also been observed for the cluster in the canonical configuration (see the right panel of Fig. 2.6). In the context of this section, we thus argue that they result from the Kozai-Lidov dynamics of the stars from the cluster in the potential of the disc, forming just after the damping effect of the spherical mean potential of the cluster is sufficiently disturbed by the ongoing formation of the cluster ***e***-structure. On the other hand, the formation of the ***e***-structure gradually decreases the significance of the overabundances and, eventually, the overabundances are entirely absorbed by the ***e***-structure.

2.3.2 Formation of the e-Structure in the Cluster

Within the K0a-model, we have investigated the orbital evolution of the initially spherically symmetric star cluster under the assumption that the masses of its individual stars are negligibly small. Here, we consider the stars to have a non-negligible mass, $m_c \sim 10^{-5} M_\bullet$. In doing so, we include their mutual gravitational interaction (hereafter the K0b-model; see Table 2.1).

The results of our calculations within the K0b-model are demonstrated in Fig. 2.17 which displays the initial (left panel) and evolved (right panel) distribution of the eccentricity vector angles Ω_e (abscissa) and i_e (ordinate) of the orbits in the cluster. We see that, while the initial distribution corresponds to the initial spherical symmetry of the cluster, in the evolved state, the orbits form a significant $\boldsymbol{e}$-structure. It thus appears that the $\boldsymbol{e}$-instability is a generic process which occurs in star clusters that are exposed to an axisymmetric perturbation, even if the perturbative potential is predefined (see also Sect. 2.3.4).

Our calculations further show that the orbital properties of this $\boldsymbol{e}$-structure are similar to those of the $\boldsymbol{e}$-structure observed in the canonical configuration. In particular, the $\boldsymbol{e}$-structure slowly rotates around the symmetry axis of the disc and, at the time of its formation, it consists of mostly low-eccentric orbits whose inclinations are distributed similar to the initial state of the cluster, showing the mean value of $\approx\pi/2$ (see Fig. 2.18). Like in the case of the canonical configuration, we attribute this to the damping effect of the mean potential of the cluster ($N_c m_c = 0.15\,M_r$) which suppresses the Kozai-Lidov mechanism in the potential of the ring, as long as its spherical symmetry is not disturbed by the formation of the $\boldsymbol{e}$-structure.

2.3.3 Kozai-Lidov Dynamics in the Disc

Within the canonical configuration, the $\boldsymbol{e}$-structures in the evolved cluster form a rather complex system (see Sect. 2.2.6). In order to describe their impact on the embedded disc, we attempt to simplify the problem by neglecting the gravity of all but the innermost $\boldsymbol{e}$-structure. This approximation is well justified by the following argument. The $\boldsymbol{e}$-structures in the cluster affect the disc on the Kozai-Lidov timescale $T_{\mathrm{KL}} \propto R_{\boldsymbol{e}\mathrm{str}}^3/M_{\boldsymbol{e}\mathrm{str}}$, where $M_{\boldsymbol{e}\mathrm{str}}$ and $R_{\boldsymbol{e}\mathrm{str}}$ denote the mass and the characteristic radius of the $\boldsymbol{e}$-structure, respectively. Since the initial radial density profile of the cluster is, in the canonical configuration, given by (1.30), the mass of the $\boldsymbol{e}$-structure can be estimated as $M_{\boldsymbol{e}\mathrm{str}} \propto R_{\boldsymbol{e}\mathrm{str}}^{1/4}$. Consequently, we find that the Kozai-Lidov timescale grows steeply as $T_{\mathrm{KL}} \propto R_{\boldsymbol{e}\mathrm{str}}^{11/4}$. Due to the assumed surface density profile of the disc, $\Sigma(R) \propto R^{-2}$, majority of the stars in the disc are located in its inner parts, i.e. in the immediate vicinity of the innermost $\boldsymbol{e}$-structure in the cluster. Hence, it appears that the innermost $\boldsymbol{e}$-structure represents the dominating perturbation of the stellar motions in the disc since the Kozai-Lidov timescale for this $\boldsymbol{e}$-structure is much shorter than those of the $\boldsymbol{e}$-structures at larger radii.

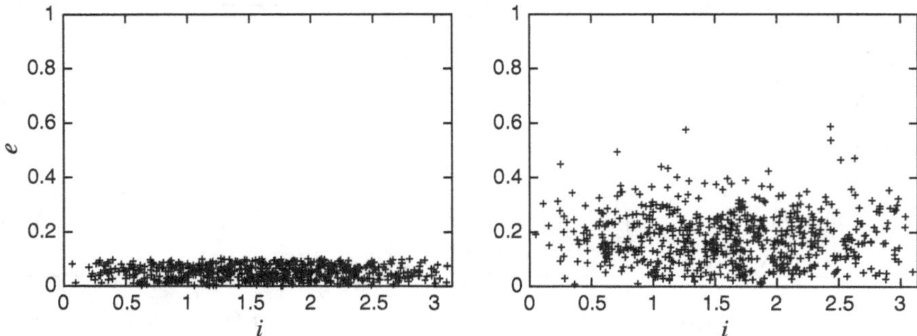

Fig. 2.18 Eccentricity e and inclination i of all the orbits in the cluster in its initial state (*left panel*) and shortly after the formation of a significant $\boldsymbol{e}$-structure ($t \approx 2.4 \times 10^3\ T_{\rm orb}(R_{\rm in})$; right panel; see also Fig. 2.17). We observe that the orbits in the $\boldsymbol{e}$-structure are still rather low-eccentric with inclinations close to their initial values. Model parameters correspond to the K0b-model (see Table 2.1)

Furthermore, as the rotation of the $\boldsymbol{e}$-structures around the symmetry axis of the disc is slow (see Sect. 2.2.2), we neglect this feature and approximate the $\boldsymbol{e}$-structure by a static predefined potential, in particular, by the potential of an infinitesimally thin ring that is centred on the SMBH and perpendicular to the plane of the disc. For simplicity, we also exclude the residual spherically symmetric component of the potential of the cluster. Moreover, we consider the stars in the disc to be test particles, thus neglecting their mutual gravitational interaction.

In summary, we focus on the model with the following three components (hereafter K1a-model, see Table 2.1):

- the central SMBH represented by the Keplerian potential,
- the stellar disc formed by test particles,
- the innermost $\boldsymbol{e}$-structure in the cluster modelled by the predefined potential of an infinitesimally thin ring which is perpendicular to the plane of the disc.

The orbital evolution of the disc within the K1a-model is followed numerically, by means of the Mbody code. The xy-plane of our Cartesian reference frame is defined as the plane of symmetry of the disc. The direction of the z-axis is determined by the initial mean angular momentum of the orbits in the disc, i.e. the stellar motions in the disc are considered to be initially prograde.

As we have learned in Sect. 1.3, the potential of the ring causes the Kozai-Lidov oscillations of eccentricity and inclination with the combined nodal and apsidal precession of the orbits in the disc. Provided the reference plane is identified with the plane of the ring, the orbital elements (e', i', Ω', ω') follow Eqs. (1.12). An example of such an evolution is displayed in Fig. 1.3. However, since we evaluate the orbital elements in the reference plane which is identified with the plane of the disc and, therefore, perpendicular to the ring, the evolution of the elements that describe the spatial orientation of the orbits (i, Ω, ω) is different. On the other hand, evolution of e remains untouched as this element does not depend upon the coordinate system.

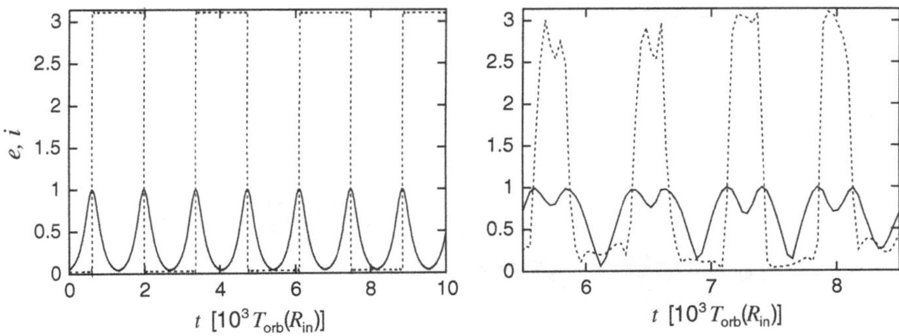

Fig. 2.19 Evolution of eccentricity e (*solid lines*) and inclination i (*dashed lines*) of individual stellar orbits from the disc within the K1a-model (*left panel*) and the canonical configuration (*right panel*; cut from the evolution displayed in the bottom panel of Fig. 2.8). We see that both orbits undergo extreme oscillations during which the maximum eccentricity, $e \to 1$, is reached for $i \approx \pi/2$. Model parameters are summarized in Table 2.1

The left panel of Fig. 2.19 shows the evolution of eccentricity e (solid line) and inclination i (dashed line) of a stellar orbit from the disc within the K1a-model. For the sake of clarity, the chosen orbit is initially almost exactly perpendicular to the plane of the ring: $i'_0 \to \pi/2$. In such a case, the evolution of i' and Ω' leads to rectangular-shaped oscillations of inclination i over the interval $\langle 0; \pi \rangle$. For orbits that are initially less inclined with respect to the ring, the oscillations of i show periodic variations of their amplitude. Unlike inclination, we see that eccentricity e of the orbit evolves in the same way as we observed in Fig. 1.3. Let us also mention that the maximum of eccentricity, $e \to 1$, is reached whenever the orbit becomes perpendicular to the parent disc, $i = \pi/2$, in contrary to what we can see in Fig. 1.3.

Evolution of the root-mean-square eccentricity, $e_{\rm rms}$, and inclination, $i_{\rm rms}$, of all the orbits in the disc within the K1a-model is displayed in the top panels of Fig. 2.20. We observe that the evolution of both elements consists of three qualitatively different stages: (i) the initial steady phase, (ii) the rapid increase and (iii) the saturated phase of damped oscillations. It turns out that this behaviour is a straightforward consequence of the averaging over a number of individual orbits that undergo the above-described oscillations. In particular, within the K1a-model, the disc is considered to be initially thin and near-circular. Hence, both eccentricities and inclinations of the individual orbits are initially low. As demonstrated in the left panel of Fig. 2.19, the orbits start their evolution in the phase when both $\mathrm{d}e/\mathrm{d}t \approx 0$ and $\mathrm{d}i/\mathrm{d}t \approx 0$. Consequently, also the root-mean-square values of the two elements are, in the initial stage of their evolution, roughly constant. In the case of eccentricity, however, this phase is very short (the rapid increase starts at about $t \approx 10^2\, T_{\rm orb}(R_{\rm in})$; top-left panel of Fig. 2.20) as the minima of its oscillations are sharp. On the other hand, the initial steady phase of the evolution of the root-mean-square inclination is, due to the rectangular-shaped oscillations, notably longer (until $t \approx 4 \times 10^2\, T_{\rm orb}(R_{\rm in})$; top-right panel of Fig. 2.20).

Subsequently, as the eccentricity and inclination of the individual orbits in the disc start to oscillate, the root-mean-square values of both elements increase. The rate of

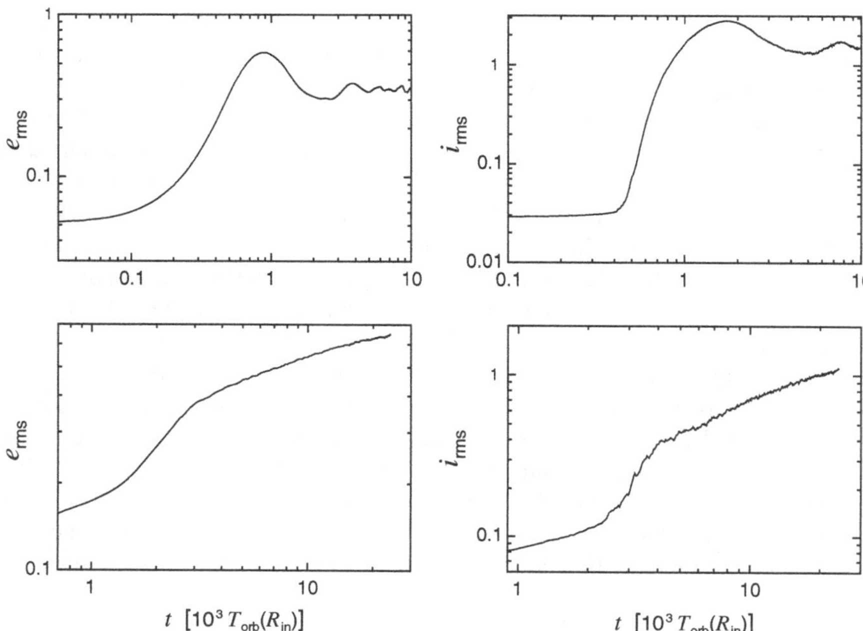

Fig. 2.20 Evolution of the root-mean-square eccentricity $e_{\rm rms}$ (*left panels*) and inclination $i_{\rm rms}$ (*right panels*) of the orbits in the disc within the K1a-model (*top panels*; results averaged over 10 realizations) and the canonical configuration (*bottom panels*; redrawn from Fig. 2.4; results averaged over 9 realizations). We see that the evolution of both root-mean-square elements within the simple K1a-model includes the accelerated phase, in accord with the results obtained for the canonical configuration. Due to high numerical demands of the problem, we have not been able to reach, in the canonical configuration, the saturated phase of the evolution. Model parameters are summarized in Table 2.1

this increase and the overall shape of the corresponding curve are strongly influenced by the assumed radial density profile of the disc which determines the distribution of the orbital semi-major axes and, consequently, also the periods ($\sim T_{\rm KL} \propto a^{3/2}$) of the oscillations. In the case of our K1a-model, majority of the orbits are located near the inner edge of the disc, thus having similar semi-major axes. Hence, eccentricity and inclination oscillate with similar frequency for most of the orbits in the disc, which leads to the observed rapid increase of the root-mean-square values of both elements.

Finally, when most of the orbits are oscillating, the root-mean-square values of both elements saturate, further showing only low-amplitude, gradually damped oscillations. The particular shape and amplitude of these oscillations are determined by the radial density profile of the disc. Hence, within the K1a-model, the similar semi-major axes of the orbits in the disc lead to the first smooth bump on the corresponding curves. Later on, as the phases of the oscillations of the individual orbits in the disc disperse, the oscillations of the root-mean-square elements slowly diminish, leaving them roughly constant.

In the context of these findings, we suggest that the accelerated evolution of the root-mean-square eccentricity and inclination of the orbits in the disc observed in the canonical configuration (redrawn in the bottom panels of Fig. 2.20) is, indeed, caused by averaging over an increasing number of oscillating orbits in the flattened potential of the innermost $\boldsymbol{e}$-structure in the cluster. This conclusion is supported by comparison of the evolution of the typical orbits from the central parts of the disc in the canonical configuration and the K1a-model which is shown in Fig. 2.19. We see that both orbits undergo similar oscillations of their eccentricity and inclination. In particular, the oscillations have nearly the same amplitude and fulfil the condition that $e \to 1$ for $i \approx \pi/2$. On the other hand, when the orbit from the canonical disc reaches the retrograde inclination $i \approx \pi$, its eccentricity does not decrease to zero, unlike the orbit from the K1a-model. Similarly, the curves that describe the evolution of the root-mean-square elements in the two configurations are also somewhat different (see Fig. 2.20). We attribute these differences to the approximations introduced in the K1a-model. Most notably, the innermost $\boldsymbol{e}$-structure in the cluster in the canonical configuration is not geometrically thin and it is developed gradually, in contrast to the predefined thin ring considered in the K1a-model. Furthermore, while the stars in the canonical configuration are gravitating, in the K1a-model, they are treated as test particles.

2.3.4 Formation of the e-Structure in the Disc

The top panels of Fig. 2.21 display the comparison of the initial (left panel) and evolved distribution (right panel) of the eccentricity vector angles Ω_e and i_e of the orbits in the disc within the K1a-model. As we can see, the initial distribution is in agreement with the assumption that the disc is initially thin and axially symmetric. In the evolved state, however, we observe that the individual eccentricity vectors are concentrated in four roughly equally populated compact groups which are uniformly distributed along the equatorial line of the plot with the step of $\Delta\Omega_e \approx \pi/2$. Furthermore, we find that these groups are not rotating in any direction. On the other hand, they show a continuous exchange of the individual orbits that freely migrate among them (see the outliers in the top-right panel of Fig. 2.21).

The picture changes if the disc is considered to be self-gravitating, i.e. if the mutual gravitational interaction of the individual stars in the disc is included (hereafter K1b-model, see Table 2.1). In particular, as we observe in the bottom-left panel of Fig. 2.21, the four groups are replaced by a single $\boldsymbol{e}$-structure that includes most of the orbits from the disc. Our calculations further show that this $\boldsymbol{e}$-structure slowly rotates around the initial symmetry axis of the disc and consists of mostly eccentric and low-inclined orbits (see the left panel of Fig. 2.22).

A similar $\boldsymbol{e}$-structure developed in the disc within the canonical configuration (see the bottom-right panel of Fig. 2.21 and the right panel of Fig. 2.22). Hence, based on the results presented in this section, we suggest that the $\boldsymbol{e}$-instability in the disc in the canonical configuration is, indeed, due to the flattened potential of the innermost

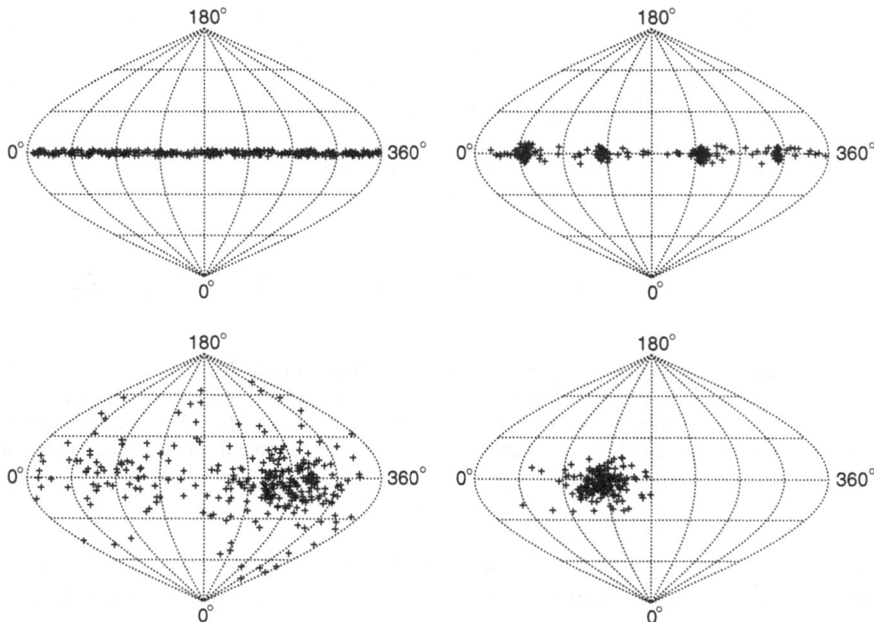

Fig. 2.21 Directions of the eccentricity vectors of the orbits in the disc in terms of angles Ω_e (abscissa) and i_e (ordinate): the initial state (*top-left panel*; all orbits in the disc are displayed) and the evolved state within the K1a-model ($t \approx 3 \times 10^3\, T_{\rm orb}(R_{\rm in})$; *top-right panel*; all orbits displayed), the evolved state within the K1b-model ($t \approx 10 \times 10^3\, T_{\rm orb}(R_{\rm in})$; *bottom-left panel*; all orbits displayed) and the evolved state in the canonical configuration ($t \approx 6 \times 10^3\, T_{\rm orb}(R_{\rm in})$; *bottom-right panel*, redrawn from Fig. 2.9; only orbits with the osculating semi-major axes $a < 1.5\, R_{\rm in}$ are shown). We see that the self-gravitating disc in the predefined potential of thin ring (K1b-model) forms a significant $\boldsymbol{e}$-structure, similar to the disc in the canonical configuration. Model parameters are summarized in Table 2.1

$\boldsymbol{e}$-structure in the cluster. Let us note that the $\boldsymbol{e}$-structure in the disc appears to be formed, in both configurations, with a roughly perpendicular orientation ($\Delta\Omega_e \approx \pi/2$) to the plane of symmetry of the perturbing potential: the infinitesimally thin ring in the case of the K1b-model and the innermost $\boldsymbol{e}$-structure in the cluster in the canonical configuration.

2.4 Discussion of Model Parameters

In the canonical configuration, we have treated the spherical cluster as a group of $N_{\rm c} = 1.25 \times 10^4$ gravitating particles. In order to investigate the impact of the number of particles in the cluster upon the acquired results, we have performed two additional sets of calculations with $N_{\rm c} = 5 \times 10^4$ and 6.25×10^3 (see also Table 2.1, models A and D, respectively). The total mass of the cluster in these models is

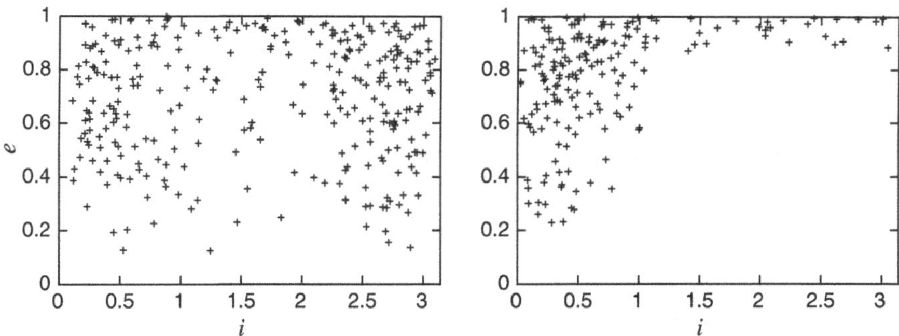

Fig. 2.22 Eccentricity e and inclination i of all the orbits in the evolved disc within the K1b-model (*left panel*) and of the innermost orbits (with the osculating semi-major axes $a < 1.5\,R_{\rm in}$) from the disc in the canonical configuration (*right panel*; redrawn from Fig. 2.10). We see that the orbits are mostly low-inclined and highly eccentric in both cases. Model parameters are summarized in Table 2.1

equal to the canonical one, i.e. the masses of the individual particles are smaller (larger). Evolution of the root-mean-square eccentricity, $e_{\rm rms}$, and inclination, $i_{\rm rms}$, in comparison to the canonical configuration is shown in the top panels of Fig. 2.23. We see that the corresponding curves are nearly identical in all three cases. This indicates that, while the treatment of the cluster gravity affects the evolution of the disc significantly (analytic potential vs. particles), the concrete number of particles in the cluster is of no importance.

On the other hand, when the total mass of the cluster is assumed to be ten times larger than the canonical one (see Table 2.1, model H), our calculations show that the accelerated evolution of both root-mean-square elements onsets by a factor of 3–4 sooner and also the rate of the growth is noticeably higher (bottom panels of Fig. 2.23, dashed lines). We attribute these results to higher masses of the $\boldsymbol{e}$-structures that form in the cluster which, therefore, cause a faster evolution of the orbits in the disc.

The solid lines in the bottom panels of Fig. 2.23 describe the case when we have included, beside the same spherical cluster of particles as within model D, an additional predefined analytic spherically symmetric potential (see Table 2.1, model M). Its radial profile and strength parameter have been considered to correspond to the matter distribution in the cluster within model D. We observe that none of the root-mean-square elements undergoes the accelerated phase of its evolution, which is likely to be due to the damping effect of the predefined analytic spherical potential that suppresses the Kozai-Lidov mechanism in the potential of the $\boldsymbol{e}$-structures in the cluster. On the other hand, our results indicate that the $\boldsymbol{e}$-structure formation neither in the cluster nor in the disc is affected significantly by the additional potential, except for the fact that both $\boldsymbol{e}$-structures appear to be formed with the same initial orientation (see Fig. 2.24).

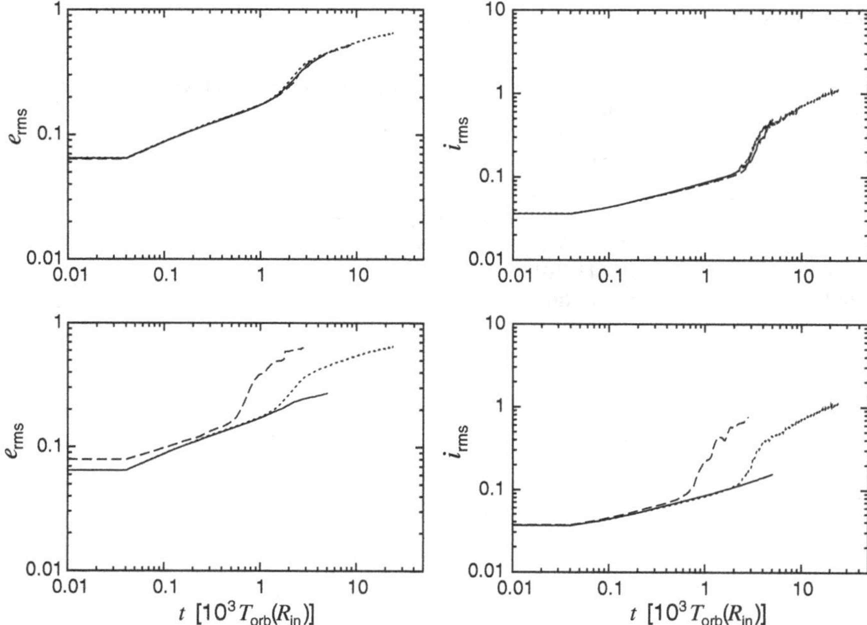

Fig. 2.23 Evolution of the root-mean-square eccentricity $e_{\rm rms}$ (*left panels*) and inclination $i_{\rm rms}$ (*right panels*) of the orbits in the disc in various configurations in comparison to the results acquired in the canonical configuration (depicted by the *dotted lines*; redrawn from Fig. 2.4; results averaged over 9 realizations). The *top panels* demonstrate the effect of the number of particles, $N_{\rm c}$, in the cluster whose total mass is kept unchanged: $N_{\rm c} = 5 \times 10^4$ (*dashed lines*; model A; results are averaged over 5 realizations) and 6.25×10^3 (*solid lines*; model D; results are averaged over 7 realizations). The *bottom panels* show the results obtained in two cases: with the total mass of particles in the cluster, $N_{\rm c} m_{\rm c}$, ten times larger than the canonical one (*dashed lines*; model H; results averaged over 9 realizations) and with an additional predefined analytic spherically symmetric potential (*solid lines*; model M; results averaged over 11 realizations). Model parameters are summarized in Table 2.1

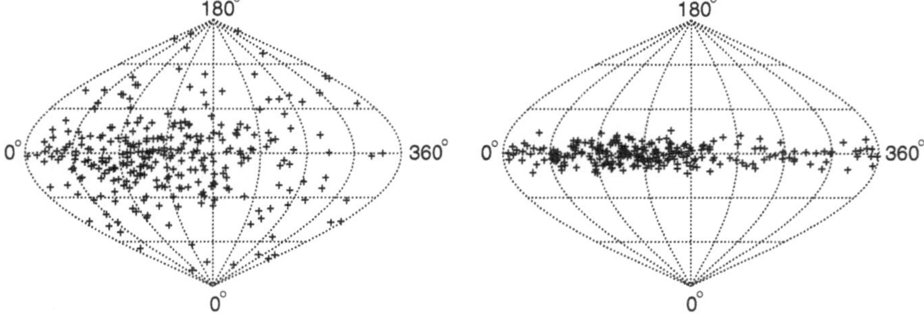

Fig. 2.24 Distribution of the directions of the eccentricity vectors of the orbits in the cluster (*left panel*) and the disc (*right panel*) at $t \approx 3.2 \times 10^3\ T_{\rm orb}(R_{\rm in})$ within model M (only orbits with the osculating semi-major axes $a < 2\,R_{\rm in}$ are displayed). We see that both the cluster and the disc evolve significant $\boldsymbol{e}$-structures. Model parameters are summarized in Table 2.1

References

Aarseth S. J., 2003, Gravitational N-Body Simulations. Cambridge Univ. Press, Cambridge
Bartko H. et al., 2009, ApJ, 697, 1741
Bartko H. et al., 2010, ApJ, 708, 834
Bender R. et al., 2005, ApJ, 631, 280
Haas J., Šubr L., 2012, JPhCS, 372, 012059
Lauer T. R., Bender R., Kormendy J., Rosenfield P., Green R. F., 2012, ApJ, 745, 121
Levin Y., Beloborodov A. M., 2003, ApJ, 590, L33
Paumard T. et al., 2006, ApJ, 643, 1011
Šubr L., 2006, private communication

Chapter 3
Dynamical Coupling of Near-Keplerian Orbits

So far, we have been investigating the dynamical evolution of a thin stellar disc which is embedded in an extended spherically symmetric star cluster, both centred on the SMBH. We have seen that the mutual interaction of these objects leads to formation of $\boldsymbol{e}$-structures in the cluster which, subsequently, significantly affect the dynamics of the stars in the disc. The investigated configuration has been primarily motivated by the observations of the Galactic Centre which show presence of both the included objects in this region: (i) stellar disc formed by massive early type stars, and (ii) extended spherical cluster of late-type stars. However, another observation of this region reveals an additional source of gravity—an axisymmetric massive molecular torus. Since similar torus is also a part of the currently widely accepted model of active galactic nuclei, the inclusion of an axisymmetric component into our previously considered model represents a natural next step in our work that shall be made in this chapter, following the paper of Haas et al. (2011). In particular, we intend to investigate the orbital evolution of a system of N mutually interacting stars on initially circular orbits around the dominating central mass under the perturbative influence of a distant axisymmetric source and an extended spherical potential. For simplicity, we focus on the case when the secular evolution of orbital eccentricities is suppressed by the spherical perturbation, which we model by a predefined analytic potential, and derive semi-analytic formulae for the evolution of normal vectors of the individual orbits.

To set the stage, we first develop a secular theory of orbital evolution for two (later in the chapter generalized to multiple) stars orbiting a massive centre, the SMBH, taking into account their mutual gravitational interaction and perturbations from the spherical stellar cluster and the axisymmetric source that is taken equivalent to an infinitesimally thin ring. It should be, however, pointed out that generalization to a more realistic structure, such as thin or thick disc, is straightforward in our setting but we believe at this stage it would just involve algebraic complexity without bringing any new quality to the model. In the same way, the stellar cluster is reduced to an equilibrium spherical model without involving generalizations beyond that level.

J. Haas, *Symmetries and Dynamics of Star Clusters*,
Springer Theses, DOI: 10.1007/978-3-319-03650-2_3,

© Springer International Publishing Switzerland 2014

For instance, an axisymmetric component of the stellar cluster may be effectively accounted for by the ring effects in the first approximation.

We are going to use standard tools of classical celestial mechanics, based on the first-order secular solution using the perturbation methods (for general discussion, see, e.g. Morbidelli 2002; Bertotti et al. 2003). In particular, the stellar orbits are described using a conventional set of the Keplerian elements which are assumed to change according to the Lagrange equations. Since we are interested in a long-term dynamical evolution of the stellar orbits, we replace, in the context of Sect. 1.2, the perturbing potential (or potential energy) with its average value over one revolution of the stars about the centre, which is the proper sense of addressing our approach as secular. As an implication of our approach, the orbital semi-major axes of the stellar orbits are constant and information about the position of the stars on their orbits is irrelevant. The secular system thus consists of description of how the remaining four orbital elements, eccentricity, inclination, longitude of node and argument of pericentre, evolve in time. This is still a very complicated problem in principle, and we shall adopt a simplifying assumption that will allow us to treat the eccentricities and pericentres separately and leave us finally with the problem of dynamical evolution of inclinations and nodes (Sect. 3.1). Let us note that this is where our approach diverges from typical applications in planetary systems, in which this separation is often impossible.

For this purpose, we consider the stellar orbits to be circular during the whole evolution of the system. Together with the assumption of well-separated orbits with constant semi-major axes, this prevents the close encounters of the stars. In this case only and under the assumption that there are no orbital mean motion resonances between the stars, the mutual interaction of the stars may be reasonably considered a perturbation to the dominating potential of the SMBH. According to Sect. 1.3, however, the assumption of circular orbits is not obviously fulfilled. In particular, initially circular orbits may be, in the axisymmetric potential, driven over the Kozai-Lidov timescale (1.13) to a very high eccentricity state due to the Kozai-Lidov oscillations. Again, this would disallow us to treat the mutual interaction of the stars as a perturbation to the dominating central potential. However, as we have learned in Sect. 1.4, this effect can be disabled by means of fast rotation of the orbital pericentres due to gravity of the extended spherically symmetric star cluster. In other words, having enough mass in the cluster may produce a strong enough perturbation to maintain small eccentricity of an initially near-circular orbit.

3.1 Evolution of Two Circular Orbits

Having stated our assumptions about semi-major axes, eccentricity and pericentre of the stellar orbits, we may now turn to description of the evolution of the two remaining orbital elements—inclination and nodal longitude. We start with a model of two interacting stars and later generalize it to the case of an arbitrary number of stars. Let us mention that the orbit-averaged potential energy (1.17) of the spherical cluster depends on the semi-major axis and eccentricity only, and thus does not influence evolution of inclination and node. For that reason it drops from our analysis in this section.

The interaction potential energy $\mathcal{R}_{\rm i}(\boldsymbol{r},\boldsymbol{r}')$ for two point sources of masses m and m' at relative positions $\boldsymbol{r}$ and $\boldsymbol{r}'$ with respect to the centre reads as

$$\mathcal{R}_{\rm i}(\boldsymbol{r},\boldsymbol{r}') = -\frac{Gmm'}{r} \sum_{l \geq 2} \alpha^l P_l\left(\cos S\right) , \tag{3.1}$$

where $P_l(x)$ are Legendre polynomials, $\cos S \equiv \boldsymbol{r} \cdot \boldsymbol{r}'/rr'$ and $\alpha \equiv r'/r$. Note that Eq. (3.1) provides the interaction energy as it appears in the equation of relative motion of stars with respect to the centre. Hence, the perturbation series start with a quadrupole term ($l = 2$). Furthermore, let us also mention that Eq. (3.1) has been used by Kozai (1962) to describe the interaction of the asteroid and Jupiter, in his setting, leading to the averaged potential (1.10).

The series on the right-hand side of Eq. (3.1) converges for $r' < r$. Since we are going to apply (3.1) to the simplified case of two circular orbits, we may replace distances r and r' with the corresponding values of semi-major axis a and a', such that $\alpha = a'/a$ now (note that the orbit whose parameters are denoted with a prime is thus assumed interior). The averaging of the interaction energy over the uniform orbital motion of the stars about the centre, implying periodic variation of S, is readily performed by using the addition theorem for spherical harmonics. This allows us to decouple unit direction vectors in the argument of the Legendre polynomial P_l and easily obtains the required average of $\mathcal{R}_{\rm i}$ over the orbital periods of the two stars. After a simple algebra we obtain

$$\overline{\mathcal{R}}_{\rm i} = -\frac{Gmm'}{a} \Psi\left(\alpha, \boldsymbol{n} \cdot \boldsymbol{n}'\right) , \tag{3.2}$$

where $\boldsymbol{n} = [\sin i \sin \Omega, -\sin i \cos \Omega, \cos i]^{\rm T}$ and $\boldsymbol{n}' = [\sin i' \sin \Omega', -\sin i' \cos \Omega', \cos i']^{\rm T}$ are unit vectors normal to the mean orbital planes of the two stars, and

$$\Psi\left(\zeta, x\right) = \sum_{l \geq 2} \left[P_l\left(0\right)\right]^2 \zeta^l P_l\left(x\right) . \tag{3.3}$$

As expected, the potential energy is only a function of: (i) the orbital semi-major axes through dependence on a and α, and (ii) the relative configuration of the two orbits in space given by the scalar product $\boldsymbol{n} \cdot \boldsymbol{n}'$. Note also that the series in (3.3) contain only even multipoles l ($P_l(0) = 0$ for l odd) and that they converge when $\zeta < 1$. However, a special care is needed when ζ is very close to unity, thus the two stellar orbits are close to each other, when hundreds to thousands of terms are needed to achieve sufficient accuracy. Still, we found it is very easy to setup an efficient computer algorithm, using recurrent relations between the Legendre polynomials, which is able to evaluate (3.3) and its derivatives. In practice, we select a required accuracy and the code truncates the series by estimating the remaining terms. In fact, since our approach neglects small eccentricity oscillations of the orbits we are anyway not allowed to set $\zeta = \alpha = a'/a$ arbitrarily close to unity. Theoretically, we should require

$$\alpha < 1 - \left(\frac{m + m'}{3M_\bullet}\right)^{1/3} , \tag{3.4}$$

by not letting the stars approach closer than the Hill radius of their mutual interaction. In the numerical examples we present below, this sets an upper limit $\alpha < 0.98$.

The formulation given above immediately provides potential energy of the star-ring interaction. In this case, the stellar orbits are always interior to the ring with symmetry axis suitably chosen as the unity vector $\boldsymbol{e}_z$ in the direction of the z-axis of our reference system. Unlike in Sect. 1.3, we restrict now to the case of circular orbit of the star but at the low computer-time expense we may include all multipole terms till specified accuracy is achieved. As a result, the orbit-averaged interaction energy with the exterior stellar orbit is given by

$$\overline{\mathcal{R}}_{\mathrm{r}} = -\frac{GmM_{\mathrm{r}}}{R_{\mathrm{r}}} \Psi\left(a/R_{\mathrm{r}}, \cos i\right) , \tag{3.5}$$

and similarly for the interior stellar orbit:

$$\overline{\mathcal{R}}_{\mathrm{r}}' = -\frac{Gm'M_{\mathrm{r}}}{R_{\mathrm{r}}} \Psi\left(a'/R_{\mathrm{r}}, \cos i'\right) . \tag{3.6}$$

The total orbit-averaged potential energy perturbing motion of the two stars is then given by superposition of the three terms:

$$\overline{\mathcal{R}} = \overline{\mathcal{R}}_{\mathrm{i}} + \overline{\mathcal{R}}_{\mathrm{r}} + \overline{\mathcal{R}}_{\mathrm{r}}' . \tag{3.7}$$

Recalling that semi-major axis values are constant, eccentricity set to zero and thus argument of pericentre undefined, we are left to study dynamics of inclination i and i' and longitude of node Ω and Ω' values. Lagrange equations provide (see, e.g. Bertotti et al. 2003)

$$\frac{\mathrm{d}\cos i}{\mathrm{d}t} = -\frac{1}{mna^2}\frac{\partial\overline{\mathcal{R}}}{\partial\Omega} , \qquad \frac{\mathrm{d}\Omega}{\mathrm{d}t} = \frac{1}{mna^2}\frac{\partial\overline{\mathcal{R}}}{\partial\cos i} , \tag{3.8}$$

$$\frac{\mathrm{d}\cos i'}{\mathrm{d}t} = -\frac{1}{m'n'a'^2}\frac{\partial\overline{\mathcal{R}}}{\partial\Omega'} , \qquad \frac{\mathrm{d}\Omega'}{\mathrm{d}t} = \frac{1}{m'n'a'^2}\frac{\partial\overline{\mathcal{R}}}{\partial\cos i'} , \tag{3.9}$$

where n and n' denote mean motion frequencies of the two stars. Note the particularly simple, quasi-Hamiltonian form of Eqs. (3.8) and (3.9), they can also be rewritten in a more compact way using the normal vectors $\boldsymbol{n}$ and $\boldsymbol{n}'$ to the respective orbit, namely

$$\frac{\mathrm{d}\boldsymbol{n}}{\mathrm{d}t} = \boldsymbol{n} \times \frac{\partial}{\partial \boldsymbol{n}} \left(\frac{\overline{\mathcal{R}}}{mna^2} \right) , \tag{3.10}$$

$$\frac{\mathrm{d}\boldsymbol{n}'}{\mathrm{d}t} = \boldsymbol{n}' \times \frac{\partial}{\partial \boldsymbol{n}'} \left(\frac{\overline{\mathcal{R}}}{m'n'a'^2} \right) . \tag{3.11}$$

Inserting here $\overline{\mathcal{R}}$ from (3.7), we finally obtain

$$\frac{\mathrm{d}\boldsymbol{n}}{\mathrm{d}t} = \omega_{\mathrm{I}} \left(\boldsymbol{n} \times \boldsymbol{n}' \right) + \omega_{\mathrm{r}} \left(\boldsymbol{n} \times \boldsymbol{e}_z \right) , \tag{3.12}$$

$$\frac{\mathrm{d}\boldsymbol{n}'}{\mathrm{d}t} = \omega_{\mathrm{I}}' \left(\boldsymbol{n}' \times \boldsymbol{n} \right) + \omega_{\mathrm{r}}' \left(\boldsymbol{n}' \times \boldsymbol{e}_z \right) , \tag{3.13}$$

where

$$\omega_{\mathrm{I}} = -n \quad \left(\frac{m'}{M_\bullet} \right) \Psi_x \left(\alpha, \boldsymbol{n} \cdot \boldsymbol{n}' \right) , \tag{3.14}$$

$$\omega_{\mathrm{I}}' = -n'\alpha \left(\frac{m}{M_\bullet} \right) \Psi_x \left(\alpha, \boldsymbol{n} \cdot \boldsymbol{n}' \right) , \tag{3.15}$$

$$\omega_{\mathrm{r}} = -n \left(\frac{M_{\mathrm{r}}}{M_\bullet} \right) \Psi_x \left(a/R_{\mathrm{r}}, n_z \right) , \tag{3.16}$$

$$\omega_{\mathrm{r}}' = -n' \left(\frac{M_{\mathrm{r}}}{M_\bullet} \right) \Psi_x \left(a'/R_{\mathrm{r}}, n_z' \right) . \tag{3.17}$$

Note that the frequencies in (3.14)–(3.17) depend on both $\boldsymbol{n}$ and $\boldsymbol{n}'$ through their presence in the argument of

$$\Psi_x(\zeta, x) \equiv \frac{\mathrm{d}}{\mathrm{d}x} \Psi(\zeta, x) , \tag{3.18}$$

which breaks the apparent simplicity of the system of Eqs. (3.12) and (3.13).

The coupled set of Eqs. (3.12) and (3.13) acquires simple solutions in two limiting cases. First, when $m = m' = 0$ (i.e. mutual interaction of stars is neglected) the two equations decouple and describe simple precession of $\boldsymbol{n}$ and $\boldsymbol{n}'$ about $\boldsymbol{e}_z$ axis of the inertial frame with frequencies $-\omega_{\mathrm{r}} \cos i$ and $-\omega_{\mathrm{r}}' \cos i'$. The sign minus of these frequencies recalls that the orbits precess in a retrograde sense when inclinations are less than 90° and vice versa. Both inclinations i and i' are constant. In the second limit, when $M_{\mathrm{r}} = 0$ (i.e. the ring is removed), Eqs. (3.12) and (3.13) obey a general integral of total angular momentum conservation

$$m\,\boldsymbol{n} + m'\alpha^{1/2}\,\boldsymbol{n}' = \boldsymbol{K} . \tag{3.19}$$

Both vectors $\boldsymbol{n}$ and $\boldsymbol{n}'$ then precess about $\boldsymbol{K}$ with the same frequency

$$\omega_{\mathrm{p}} = \frac{\omega_{\mathrm{I}}}{m'\alpha^{1/2}} \frac{m + m'\alpha^{1/2}\left(\boldsymbol{n}\cdot\boldsymbol{n}'\right)}{\sqrt{m^2 + m'^2\alpha + 2mm'\alpha^{1/2}\left(\boldsymbol{n}\cdot\boldsymbol{n}'\right)}}\,, \tag{3.20}$$

keeping the same mutual configuration. In particular, initially coplanar orbits (i.e. $\boldsymbol{n}$ and $\boldsymbol{n}'$ parallel) would not evolve, which is in agreement with intuition.

Unfortunately, we were not able to find analytical solution to the (3.12) and (3.13) system except for these two situations described above. Obviously, it can be always solved using numerical methods as we shall discuss in Sect. 3.1.2.

3.1.1 Integrals of Motion

In general, Eqs. (3.12) and (3.13) have only two first integrals. Our assumptions about the circumnuclear torus mass distribution still provide a symmetry vector $\boldsymbol{e}_z$. Thus, while the total angular momentum $\boldsymbol{K}$ is no more conserved now, its projection onto $\boldsymbol{e}_z$ is still an integral of motion

$$m\cos i + m'\alpha^{1/2}\cos i' = C_1 = K_z. \tag{3.21}$$

Because m, m' and α are constant, Eq. (3.21) provides a direct constraint of how the two inclinations i and i' evolve. In particular, one can be expressed as a function of the other.

The quasi-Hamiltonian form of Eqs. (3.8) and (3.9) readily results in a second integral of motion

$$\overline{\mathcal{R}}\left(\cos i, \cos i', \boldsymbol{n}\cdot\boldsymbol{n}'\right) = C_2. \tag{3.22}$$

The list of arguments in $\overline{\mathcal{R}}$, as explicitly provided above, reminds that it actually depends on: (i) the inclination values i and i', and (ii) the difference $\Delta\Omega = \Omega - \Omega'$ of the nodal longitudes of the two interacting orbits. Using (3.21), the conservation of $\overline{\mathcal{R}}$ thus provides a constraint between the evolution of i and $\Delta\Omega$ (say). While not giving a solution of the problem, the constraint due to combination of first integrals (3.21) and (3.22) can still provide useful insights.

Figure 3.1 illustrates how the first integrals help understanding several features of the orbital evolution for two interacting stars at distances $a' = 0.04\,R_{\mathrm{r}}$ and $a = 0.05\,R_{\mathrm{r}}$. For the sake of simplicity we also assume their mass is equal, hence $m' = m$, and the ring has been given mass $M_{\mathrm{r}} = 0.3\,M_{\bullet}$. Data in this figure show constrained evolution of orbital inclinations i' (solid lines) and i (dashed lines) as a function of nodal difference $\Delta\Omega$. The two orbits were assumed to be initially coplanar ($\Delta\Omega = 0°$) with an inclination of $i' = i = 70°$. A set of curves correspond to different values of stellar masses, from small (1) to larger values (4), which basically means increasing strength of their mutual gravitational interaction.

First, conservation of the $\boldsymbol{e}_z$-projected orbital angular momentum, as given by Eq. (3.21), requires that increase in i' is compensated by decrease of i. This results

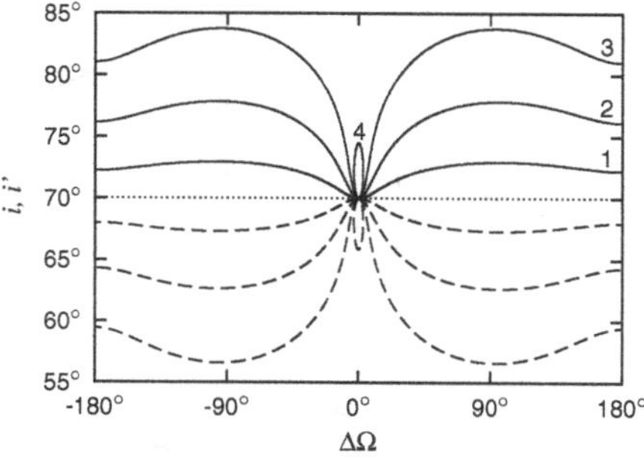

Fig. 3.1 Isolines of the $\overline{\mathcal{R}} = C_2$ integral in the i or i' versus $\Delta\Omega$ space. For the sake of example we use orbits of two equal-mass stars ($m' = m$) with semi-major axes $a' = 0.04\,R_{\rm r}$ and $a = 0.05\,R_{\rm r}$. The mass of the ring is set to $M_{\rm r} = 0.3\,M_\bullet$. The individual lines correspond to different values of stellar mass: $m = 5 \times 10^{-7}\,M_\bullet$ (*curves 1*), $m = 2 \times 10^{-6}\,M_\bullet$ (*curves 2*), $m = 5 \times 10^{-6}\,M_\bullet$ (*curves 3*), and $m = 9 \times 10^{-6}\,M_\bullet$ (*curves 4*). Both orbits have been given 70° inclination at $\Delta\Omega = 0°$ (i.e. initially coplanar and inclined orbits). *Solid lines* show inclination i' of the *inner orbit*, 'the mirror-imaged' *dashed lines* describe inclination i of the *outer orbit*

in a near-mirror-imaged evolution of the two inclinations. Using the first equation of (3.8), one finds

$$\frac{\mathrm{d}i}{\mathrm{d}t} = \frac{n}{\sin i}\,\frac{m'}{M_\bullet}\,\sin\left(\Omega - \Omega'\right)\,\Psi_x\left(\alpha, \boldsymbol{n}\cdot\boldsymbol{n}'\right)\,, \tag{3.23}$$

which straightforwardly implies that the outer stellar orbit is initially torqued to decrease its inclination while the inner orbit increases its inclination. This is because initially $\boldsymbol{n}\cdot\boldsymbol{n}' \approx 1$, and $\Psi_x(\alpha, 1)$ is positive, and at the same time, precession of the nodes is dominated by interaction with the ring which makes the outward orbit node to drift faster (and hence $\Omega - \Omega'$ is negative).

Second, Fig. 3.1 indicates there is important change in topology of the isolines $\overline{\mathcal{R}} = C_2$ as the stellar masses overpass some critical value (about $8.5 \times 10^{-6}\,M_\bullet$ in our example). For low-mass stars their mutual gravitational interaction is weak letting the effects of the ring dominate (curve 1). The orbits regularly precess with different frequency, given their different distance from the centre, and thus $\Delta\Omega$ acquires all values between $-180°$ and $180°$. The mutual stellar interaction produces only small inclination oscillation. As the stellar masses increase (curves 2 and 3) the inclination perturbation becomes larger. For super-critical values of m (curve 4) the isolines of constant $\overline{\mathcal{R}}$ become only small loops about the origin. This means that $\Delta\Omega$ is bound to oscillate in a small interval near origin and inclination perturbation becomes strongly damped. Put in words, the gravitational coupling between the stars became strong enough to tightly couple the two orbits together. Note that they still collectively precess in space due to the influence of the ring.

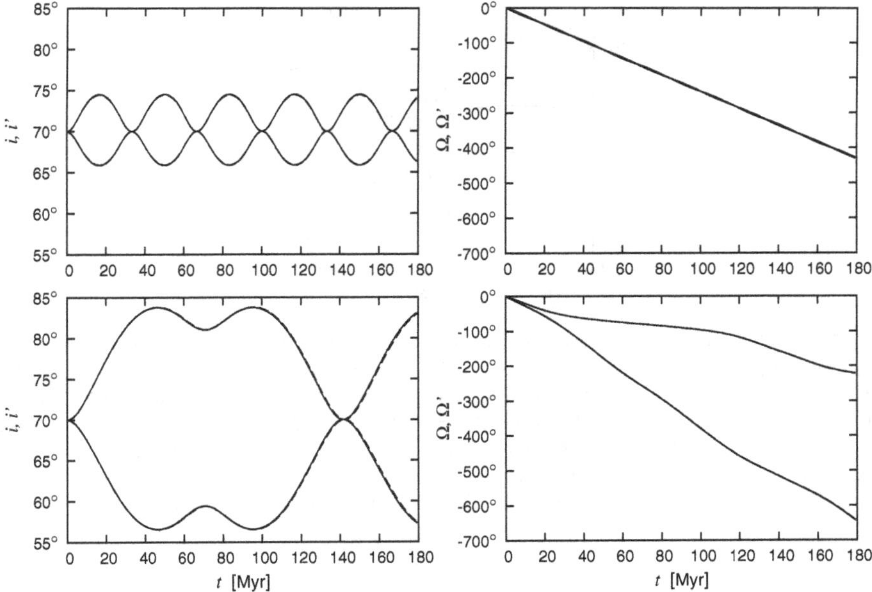

Fig. 3.2 Evolution of the system of two stars in the compound potential of the central SMBH, spherical cluster and axisymmetric ring. *Solid lines* represent solution of two-body Eqs. (3.12) and (3.13), while the *dashed lines* show result of the direct numerical integration of the equations of motion. In each panel, *upper* and *lower lines* correspond to the *inner* and *outer star*, respectively. Common parameters for both examples are the same as in Fig. 3.1; in the *upper panels*, we set $m = m' = 9 \times 10^{-6}\, M_\bullet$, while in the *lower* ones $m = m' = 5 \times 10^{-6}\, M_\bullet$

3.1.2 Numerical Solutions

In order to solve Eqs. (3.12) and (3.13) numerically, we adopt a simple adaptive step-size 4.5th-order Runge-Kutta algorithm. Let us mention that our implementation of this algorithm conserves the value of both integrals of motion C_1 and C_2 with relative accuracy better than 10^{-6}.

Two sample solutions are shown in Fig. 3.2. The upper panels represent evolution of two orbits with coupled precession which corresponds to the curve 4 in Fig. 3.1, while in the bottom panels we consider the case of lower-mass stars, whose orbits precess independently. This later mode corresponds to the curve 3 in Fig. 3.1. Besides the solution of the equations for mean orbital elements, we also show results of a full-fledged numerical integration of the particular configuration in the space of classical positions and momenta $(\boldsymbol{r}, \boldsymbol{r}'; \boldsymbol{p}, \boldsymbol{p}')$. Both solutions are nearly identical, which confirms validity of the presented secular perturbation theory.

For sake of further discussion, we find it useful to comment in a little more detail on the case of two, nearly independently precessing orbits (bottom panels on Fig. 3.2). In this case, the precession frequencies of the outer and inner star orbits are given by

ω_r and ω_r' in Eqs. (3.16) and (3.17). When truncated to the quadrupole ($l = 2$) level, sufficient for the small value of a/R_r, one has for the outer star orbit

$$\frac{d\Omega}{dt} \simeq -\frac{3}{4}\frac{\cos i}{T_{KL}} , \tag{3.24}$$

where T_{KL} is given by (1.13). A similar formula holds for the inner star orbit denoted with primed variables. As seen in Fig. 3.1, and understood from the analysis of integrals of motion in Sect. 3.1.1, the period of the evolution of the system of the two orbits is given implicitly by the difference in their precession rate: $\Omega(T_{char}) - \Omega'(T_{char}) = 2\pi$. The secular rate of nodal precession in (3.24) is not constant because the mutual gravitational interaction of the stars makes their orbital inclinations oscillate. However, in the zero approximation we may replace them with their initial values, $i = i' = i_0$ which gives an order of magnitude estimate

$$T_{char} \simeq \frac{8\pi}{3\cos i_0}\left[\frac{1}{T_{KL}} - \frac{1}{T'_{KL}}\right]^{-1} . \tag{3.25}$$

For the solution shown in the lower panels of Fig. 3.2, formula (3.25) gives $T_{char} \approx$ 460 Myr, in a reasonable agreement with the observed period of $\approx$140 Myr. When the orbital evolution is known (being integrated numerically), more accurate estimate can be obtained considering mean values of the inclinations

$$T_{char} \simeq \frac{8\pi}{3}\left[\frac{\cos\bar{i}}{T_{KL}} - \frac{\cos\bar{i}'}{T'_{KL}}\right]^{-1} . \tag{3.26}$$

For the case of the solution of the lower panel of Fig. 3.2, with $\bar{i} \approx 60°$ and $\bar{i}' \approx 80°$, formula (3.26) gives $T_{char} \approx 120$ Myr.

3.2 Generalization for Multiple Orbits

The previous formulation straightforwardly generalizes to the case of N stars orbiting the centre on circular orbits with semi-major axes a_k ($k = 1, \ldots, N$). This is because the potential energies of all pairwise interactions build the total

$$\overline{\mathcal{R}}_i = -\frac{1}{2}\sum_{k\neq l}\frac{Gm_k m_l}{a_{kl}}\,\Psi\left(\alpha_{kl}, \boldsymbol{n}_k\cdot\boldsymbol{n}_l\right) , \tag{3.27}$$

where m_k is the mass of the kth star, $a_{kl} = \min(a_k, a_l)$, $\alpha_{kl} = \min(a_k, a_l)/\max(a_k, a_l)$ and $\boldsymbol{n}_k$ is the normal vector to the orbital plane of the kth star. Similarly, interaction with the ring is simply given by

$$\overline{\mathcal{R}}_{\mathrm{r}} = -\sum_k \frac{G m_k M_{\mathrm{r}}}{a_k} \, \Psi\left(a_k/R_{\mathrm{r}}, \boldsymbol{n}_k \cdot \boldsymbol{e}_z\right) . \tag{3.28}$$

The total potential energy of perturbing interactions is

$$\overline{\mathcal{R}} = \overline{\mathcal{R}}_{\mathrm{i}} + \overline{\mathcal{R}}_{\mathrm{r}} , \tag{3.29}$$

and the equations of orbital evolution now read as

$$\frac{\mathrm{d}\boldsymbol{n}_k}{\mathrm{d}t} = \boldsymbol{n}_k \times \frac{\partial}{\partial \boldsymbol{n}_k}\left(\frac{\overline{\mathcal{R}}}{m_k n_k a_k^2}\right), \tag{3.30}$$

for $k = 1, \ldots, N$ (n_k is the frequency of the unperturbed mean motion of the k-th star about the centre). Their first integrals then can be written as

$$\sum_k m_k n_k a_k^2 \left(\boldsymbol{n}_k \cdot \boldsymbol{e}_z\right) = C_1 = K_z \tag{3.31}$$

and

$$\overline{\mathcal{R}} = C_2 . \tag{3.32}$$

Due to mutual interaction of multiple stars, solutions of Eqs. (3.30) represent, in general, an intricate orbital evolution, whose course is hardly predictable as it strongly depends upon the initial setup. Our numerical experiments show, however, that it is still possible to identify several qualitative features which remain widely valid. For instance, a group of orbits with small separations may orbitaly couple together and effectively act as a single orbit in interaction with the rest of the stellar system.

This is demonstrated in Fig. 3.3 which shows two sample solutions of Eqs. (3.30) for a system of two such groups. For the sake of clarity, each group consists only of two orbits. Individual semi-major axes are, for both solutions, set to $a_1 = 0.0373\, R_{\mathrm{r}}$, $a_2 = 0.0408\, R_{\mathrm{r}}$, $a_3 = 0.0478\, R_{\mathrm{r}}$, $a_4 = 0.0511\, R_{\mathrm{r}}$ in order to mimic the two-orbits models from Fig. 3.2. For the same reason, all the individual masses are considered equal, $m_1 = m_2 = m_3 = m_4$, and set to $2.5 \times 10^{-6}\, M_{\bullet}$ in the lower panels, while for the upper panels we assume $4.5 \times 10^{-6}\, M_{\bullet}$. The other parameters remain identical to the case of the two-orbits models. As we can see (cf. Figs. 3.3 and 3.2), the dynamical impact of each coupled pair of orbits upon the rest of the stellar system is equivalent to the effect of the corresponding single orbit if both the total mass and semi-major axis of the pair are appropriate. The individual orbits within each pair then naturally oscillate about the single-orbit solution according to their mutual interaction. This conclusion remains valid even in more complicated systems as we shall show in the following chapter.

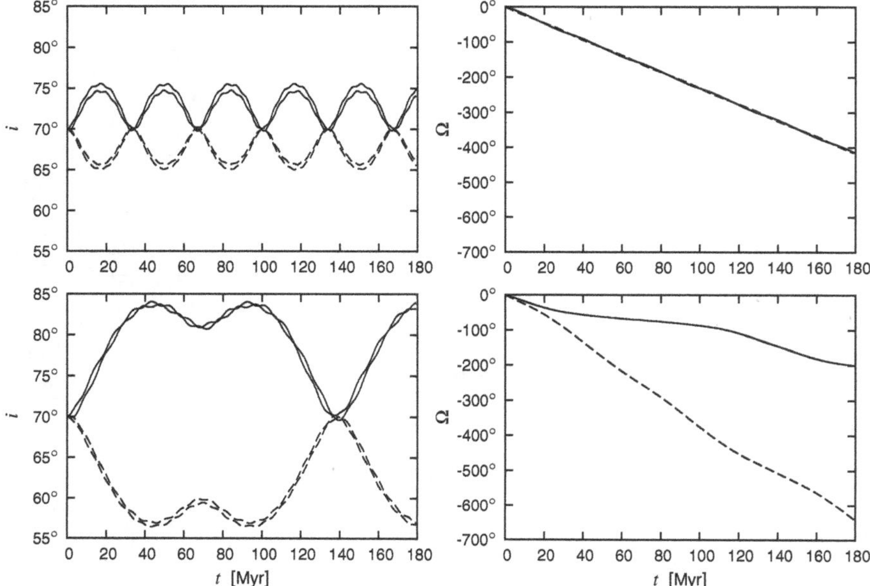

Fig. 3.3 Evolution of the system of four stars in the compound potential of the central SMBH, spherical cluster and axisymmetric ring. The stellar orbits form two couples. In both of them, the orbits have similar semi-major axes in order to mimic the system shown in Fig. 3.2. In each panel, *upper* and *lower lines* correspond to the *inner* and *outer couple*, respectively. The individual semi-major axes are for both examples set to $a_1 = 0.0373\, R_{\rm r}$, $a_2 = 0.0408\, R_{\rm r}$, $a_3 = 0.0478\, R_{\rm r}$, $a_4 = 0.0511\, R_{\rm r}$. The other common parameters for both examples are the same as in Fig. 3.1; in the *upper panels*, we set $m_1 = m_2 = m_3 = m_4 = 4.5 \times 10^{-6}\, M_{\bullet}$, while in the *lower ones* $m_1 = m_2 = m_3 = m_4 = 2.5 \times 10^{-6}\, M_{\bullet}$

References

Bertotti B., Farinella P., Vokrouhlický D., 2003, Physics of the Solar System. Kluwer Academic Publishers, Dordrecht

Haas J., Šubr L., Vokrouhlický D., 2011, MNRAS, 416, 1023

Kozai Y., 1962, AJ, 67, 591

Morbidelli A., 2002, Modern Celestial Mechanics. Taylor & Francis, London

Chapter 4
Sagittarius A*

Over the past two decades, nearly 200 early-type stars have been revealed in the innermost parsec of our Galaxy (see Genzel et al. 2010 for the most recent review; Allen et al. 1990; Genzel et al. 2003; Ghez et al. 2003, 2005; Paumard et al. 2006, Bartko et al. 2009, 2010). Observations suggest that these stars are orbiting a highly concentrated mass, which is associated with the compact radio source Sgr A*. It is widely accepted that this source is powered by a SMBH. Its mass and distance from the Sun are estimated to be approximately $4 \times 10^6\ M_\odot$ and 8 kpc, respectively (Ghez et al. 2003; Eisenhauer et al. 2005; Gillessen et al. 2009a, b; Yelda et al. 2011).

According to the most recent observations of Bartko et al. (2009, 2010), the majority (136) of the early-type stars observed in the Sgr A* region are located at projected distance 0.03 pc $\lesssim r \lesssim$ 0.5 pc from the SMBH. Roughly one half of these stars appear to form a coherently rotating disc-like structure, the so-called clockwise system (CWS; discovered by Levin and Beloborodov 2003). The remaining stars are randomly scattered off the CWS plane. Nevertheless, some authors report the existence of a second coherent structure at a large angle with respect to the CWS—the counterclockwise system (CCWS; first mentioned by Genzel et al. 2003). Even with both structures considered, however, a significant number of the early-type stars are still not belonging to either of them.

Observations have further established that all the early-type stars between 0.03 and 0.5 pc from the SMBH are either Wolf-Rayett stars or O- or early B-stars (WR/OB stars) (Bartko et al. 2009, 2010). Evolutionary phases of individual stars indicate that all of them have been formed 6 ± 2 Myr ago within a short period of time, probably not exceeding 2 Myr (Paumard et al. 2006). The presence of such stars so close to the SMBH is rather suprising. In particular, the tidal field of the SMBH is strong enough to prevent standard star formation mechanisms. Hence, various hypotheses have been suggested to explain the origin and configuration of the WR/OB stars observed in the Sgr A* region.

In situ fragmentation of a self-gravitating gaseous disc is probably the currently most widely accepted formation scenario for the stars that belong to the CWS (Levin and Beloborodov 2003; Paumard et al. 2006). This process was theoretically predicted to form stars in active galactic nuclei around SMBHs of masses 10^6–$10^{10}\ M_\odot$

J. Haas, *Symmetries and Dynamics of Star Clusters*,
Springer Theses, DOI: 10.1007/978-3-319-03650-2_4,
© Springer International Publishing Switzerland 2014

(Collin and Zahn 1999). However, as it naturally forms stars in a single disc-like structure, it fails to explain the origin of the stars observed outside the CWS. Many authors have, therefore, been seeking a mechanism that could have scattered these outliers from the parent disc plane.

It has been shown by Cuadra et al. (2008) that two-body relaxation of the parent disc does not yield the observed large inclinations of the outliers with respect to the disc plane. According to Kocsis and Tremaine (2011), some of them may have been brought to their positions by vector resonant relaxation between the disc and the cluster of late-type stars which also appears to be present in the Sgr A* region (Genzel et al. 2003; Schödel et al. 2007; Do et al. 2009). However, it is still unclear whether this process can explain the origin of the stars with line-of-sight angular momenta opposite to that of the stars within the CWS. In order to overcome this issue, Löckmann et al. (2008) have considered mutual interaction of two self-gravitating discs at large angles relative to each other. Although this mechanism indeed yields the observed configuration of the WR/OB stars, it needs rather special initial conditions. In particular, the two discs must have been formed at specific angles with respect to each other in order to stand for the CWS and CCWS. Moreover, as all the WR/OB stars seem to be coeval (Paumard et al. 2006), the two discs must have formed their stars at roughly the same time. Since this is not very likely, the need of such special initial conditions represents the major drawback of this scenario. Similar scenarios, such as the interaction of two gaseous streams (Hobbs and Nayakshin 2009), suffer from the same problem.

Šubr et al. (2009) and Šubr (2011) have suggested that all the WR/OB stars in the Sgr A* region may have been born in a single gaseous disc. They argue that the stars observed outside the CWS represent the outer parts of the parent disc, that have been partially disrupted by the gravity of the circumnuclear disc (CND). The CND is a clumpy molecular torus, that is located between 1.6 and 2.0 pc from Sgr A* (Christopher et al. 2005). The upper estimate of the total mass of this structure, which is almost perpendicular to the CWS (Paumard et al. 2006), reaches the order of $10^6\,M_\odot$. Šubr et al. (2009) claim that the gravity of the CND would cause differential precession of the individual orbits in the parent stellar disc. Such a process would force the stars from the outer parts of the disc to leave the disc plane while the inner parts of the disc would remain untouched. This core would be identified as the CWS today.

In this chapter, which follows the paper of Haas et al. (2011), we further investigate the hypothesis of Šubr et al. (2009) (see also Šubr 2011; Šubr and Haas 2012; Haas and Šubr in Proceedings of RAGtime 11/12, in press). In particular, we include the self-gravity of the parent stellar disc, and follow its orbital evolution in a predefined external potential by means of numerical N-body computations. In addition to the SMBH and the CND, the external potential includes the gravity of the cluster of late-type stars. Even though its density profile is still unclear, its potential may be considered, in the first approximation, to be spherically symmetric, and centred on the SMBH.

4.1 Introduction of the Numerical Model

Šubr et al. (2009) have considered the disc to be surrounded by the cluster of late-type stars, and the stars in the disc to be test particles. They claim that the evolution of individual stellar orbits in the disc is dominated by the Kozai-Lidov mechanism which we have discussed in Sect. 1.3. Within a simplified model where the CND is equivalent to an infinitesimally thin ring, the authors find that the Kozai oscillations become negligible when the mass, M_c, of the cluster within the radius R_{CND} of the CND fulfills the approximate condition $M_c \gtrsim 0.1\, M_{CND}$, where M_{CND} stands for the mass of the CND. In that case, the first time derivative of Ω becomes constant and can be written as (cf. Eq. (3.24))

$$\frac{d\Omega}{dt} = -\frac{3}{4}\frac{\cos i}{T_{KL}}\frac{1+\frac{3}{2}e^2}{\sqrt{1-e^2}}\,, \tag{4.1}$$

where T_{KL} is the Kozai-Lidov time-scale (1.13) and inclination, i, of the orbit is the angle between the symmetry axis of the CND and the angular momentum of the star. According to this formula, the rate of precession strongly depends upon the semi-major axis of the orbit. Hence, the outer parts of the disc are more affected by the precession than the inner parts and, therefore, the disc becomes warped or, eventually, completely disrupted.

With the gravity of the stars in the disc included, the orbital evolution of the disc may be affected by the dynamical coupling of the individual stellar orbits described in Chap. 3. Let us, therefore, further investigate the differential precession of the orbits within a self-gravitating disc. For this purpose, we introduce the model of the Galactic Centre in the following way:

- the SMBH of mass $M_\bullet = 4 \times 10^6\, M_\odot$ is considered to be a source of Keplerian potential,
- the CND is modelled as a single massive particle of mass M_{CND} orbiting the SMBH on a circular orbit of radius $R_{CND} = 1.8$ pc (see Sect. 4.3.2 for the discussion of this approximation),
- the cluster of late-type stars is represented by a smooth power-law density profile, $\rho(r) \propto r^{-\beta}$, and mass M_c within the radius R_{CND},
- the early-type stars in the disc are treated as N gravitating particles, $m \in [m_{min}, m_{max}]$, distributed according to a power-law mass function $dN \propto m^{-\alpha}dm$.

The stellar orbits in the disc are constructed to be initially geometrically circular. However, due to the additional spherically symmetric component of the gravitational potential (cluster of late-type stars), the osculating eccentricities do not truthfully describe real curvature of the orbits in space. For example, the initial osculating eccentricity of the outermost orbits in the disc is $\approx$0.1 for $M_c = 0.1$ and $\beta = 7/4$. Initial radii of the orbits are, in accord with the observations (Paumard et al. 2006; Bartko et al. 2009, 2010), generated randomly between 0.04 and 0.4 pc. Their distribution obeys $dN \propto a^{-1}da$. The disc is initially thin with half-opening angle

Table 4.1 Parameters of the canonical model (see Fig. 4.1 for the corresponding results)

$M_\bullet = 4 \times 10^6\,M_\odot$	$M_{\rm CND} = 0.3\,M_\bullet$	$M_{\rm c} = 0.03\,M_\bullet$
$m = 4$–$120\,M_\odot$	$\alpha = 1$	$\beta = 7/4$
$N = 200$	$i^0_{\rm CWS} = 70°$	$\Delta_0 = 2.5°$

Total mass of the young stellar disc $\approx 6.8 \times 10^3\,M_\odot$
($\gtrsim 90\,\%$ by ≈ 140 stars with $m \geq 12\,M_\odot$)

$\Delta_0 \lesssim 5°$. The initial inclination of the disc plane with respect to the CND, which is defined by the mean angular momentum of the stars in the disc, is denoted $i^0_{\rm CWS}$. We follow the evolution of this system numerically, by means of the N-body integration code NBODY6. The gravitational potentials of both the SMBH and the cluster of late-type stars have been incorporated into the original code as additional external fields.

4.2 Coherent Evolution of the Core of the Disc

The evolution of the stellar disc may, in principle, depend upon all the parameters of the system under consideration: $M_{\rm CND}$, $M_{\rm c}$, β, N, $m_{\rm min}$, $m_{\rm max}$, α, Δ_0, and $i^0_{\rm CWS}$. Hence, in order to investigate the evolution properly, it is necessary to cover all the reasonable values of these parameters. On the other hand, in order to demonstrate the results of our calculations, it is useful to define a 'canonical' model with the parameters set to the values listed in Table 4.1.

The observations indicate that all of the WR/OB stars have mass $m \gtrsim 12\,M_\odot$ (Paumard et al. 2006; Bartko et al. 2009, 2010). However, since it is likely that a number of undetected less massive early-type stars exist in the Sgr A* region, we consider $m \in [4\,M_\odot, 120\,M_\odot]$ in the canonical model. For a more convenient comparison with the currently available observational data, we display properties of only a subset of the stars with mass $m \geq 12\,M_\odot$ in figures.

Due to the stochastic nature of the studied system, the results should be averaged over a number of realizations with identical values of the model parameters in order to distinguish general trends from random fluctuations. For this purpose, we first considered 120 realizations of the canonical model. It has, however, turned out that the results become statistically relevant already for 12 realizations. Hence, we consider only 12 realizations of all the other models discussed in this chapter in order to shorten the necessary computational time ($\approx$5 h on 3 GHz CPU per run; $\approx$2000 runs in total). Since we attempt to explain the configuration of a specific observed system, every single realization represents a possible course of its evolution. We thus show the standard deviation for some of the key quantities for a more thorough description of the set of possible evolutions.

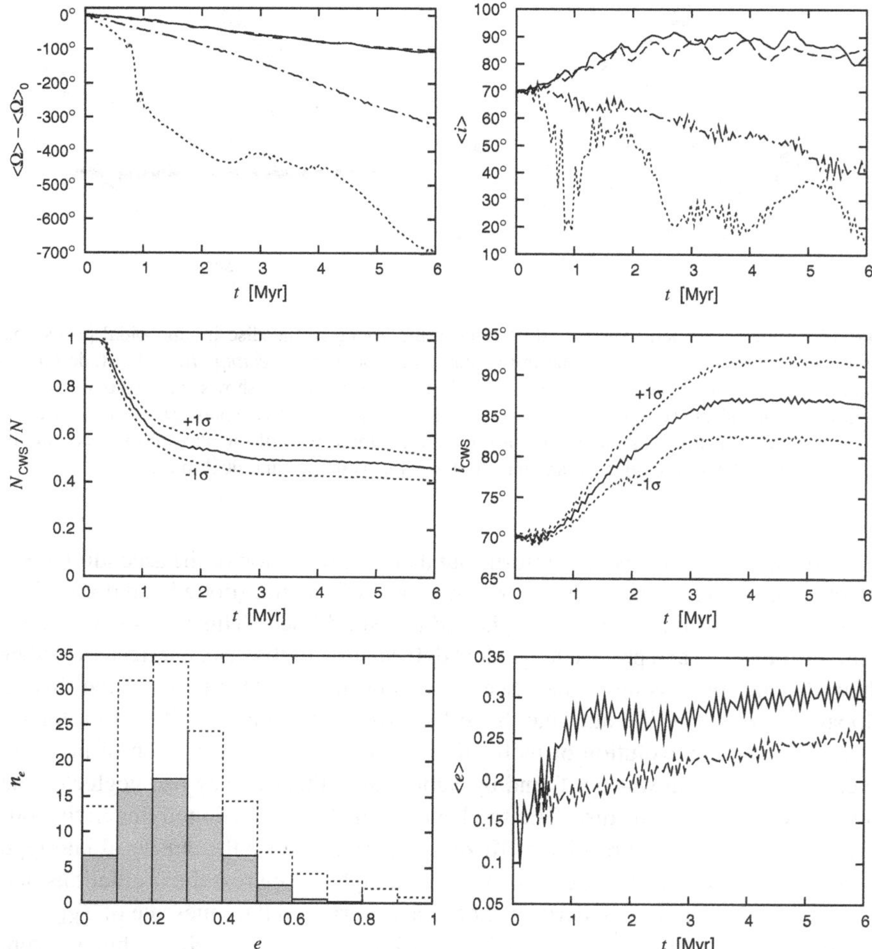

Fig. 4.1 Results for the canonical model (see Table 4.1 for the corresponding parameters). Only properties of the stars with $m \geq 12\,M_\odot$ are displayed. The *dotted lines* in the *middle panels* denote standard deviation for the set of 120 included realizations. *Top* Evolution of the mean value of Ω (*left*) and i (*right*) within different parts of the disc for one of the realizations of this model. The *solid line* describes the group of the innermost stars, followed by the *dashed, dot-dashed* and *dotted line*, which correspond to successive outer groups. *Middle-left* Number of stars within the CWS (i.e. with angular momentum deviating from the mean angular momentum of the CWS by less than 20°). *Middle-right* Inclination of the CWS with respect to the CND. *Bottom-left* Eccentricity distribution for the stars after 6 Myr of orbital evolution. The *empty boxes* denote distribution for all the stars in the young stellar system while the *grey* ones represent only stars within the CWS. *Bottom-right* Mean eccentricity of all the stars in the young stellar system (*solid line*) and within the CWS (*dashed line*)

The results for the canonical model are shown in Fig. 4.1. The top-left panel demonstrates the differential precession of the orbits in the disc. It shows the evolution of the mean value of Ω within different groups of stars, which are determined by their

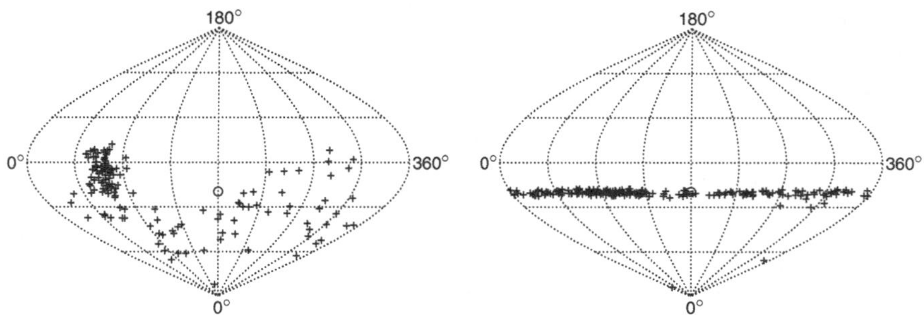

Fig. 4.2 Angular momenta of individual stars in the young stellar disc in sinusoidal projection after 6 Myr of orbital evolution. The initial state is denoted by an *empty circle*. Latitude on the plots corresponds to i while longitude is related to Ω. The *left panel* shows the results for one of the realizations of the canonical model (only stars with $m \geq 12\,M_{\odot}$ displayed). For comparison, the *right panel* illustrates the situation with negligibly small mass of the stars in the disc (single mass, $m = 0.004\,M_{\odot}$, the other parameters are the same as in the canonical model)

initial distance from the centre. It turns out that the precession of the ascending node affects more strongly the orbits in the outer parts of the disc (dotted and dot-dashed lines) than those in the inner parts (dashed and solid lines). This result is in accord with formula (4.1) and proves the gradual deformation of the disc. Our results further show that the precession of the ascending node in the outer parts of the disc is globally accelerated. We attribute this effect, which becomes significant on longer time scales, to the evolution of inclination due to two-body relaxation of the disc. Such an acceleration was not found by Šubr et al. (2009) as they had neglected the gravity of the stars in the disc. The sudden drop of $\langle\Omega\rangle$ on the dotted line in the top-left panel of Fig. 4.1 is a residue of Kozai oscillations. Since the cluster of late-type stars is, in our canonical model, not massive enough to suppress the oscillations of e and i entirely, the first time derivative of Ω also varies on the timescale of T_{KL}.

The evolution of the mean inclination $\langle i\rangle$ with respect to the CND within different parts of the disc is shown in the top-right panel of Fig. 4.1. We see that the inclination of the outer parts of the disc is decreasing (dotted and dot-dashed lines), while it grows and saturates at $\approx 90°$ in the inner parts (dashed and solid lines). We further find that the evolution of both Ω and i is similar for all the orbits in the inner parts of the disc. Hence, the core of the disc remains rather undisturbed and coherently changes its orientation towards perpendicular with respect to the CND. This effect can be seen in the left panel of Fig. 4.2, which shows the directions of angular momenta of the individual stars in the disc after 6 Myr of orbital evolution for one of the realizations of the canonical model (the initial state is denoted by an empty circle). Our results indicate that the compact group at inclination $\approx 90°$ is formed by the stars from the inner parts of the disc, while the remaining scattered stars represent the entirely dismembered outer parts. Hence, we see that the dynamical evolution of the initially thin stellar disc leads to a configuration similar to that observed in the Sgr A* region (see Paumard et al. 2006; Bartko et al. 2009, 2010). In particular, the core of the disc can be identified with the CWS observed today and, at the same time, the stars from the dissolved outer parts can stand for the WR/OB stars found outside the CWS.

In order to compare our results with the observations more thoroughly, we further define CWS within our model in the following iterative way. As the zeroth step, the CWS is considered to be formed by a fixed number of the innermost stars from the initial disc. In the next step, we exclude from the CWS all the stars whose angular momenta deviate from the mean angular momentum of the CWS by more than 20°. On the other hand, the stars initially from outside the CWS, which do not fulfill the latter condition, are included into the CWS. Then, we recalculate the mean angular momentum of the CWS and repeat the whole procedure iteratively until there are no changes of the CWS in between two subsequent steps.

Observations indicate (Paumard et al. 2006; Bartko et al. 2009, 2010) that roughly one half of the WR/OB stars are members of the observed CWS. We thus follow within our calculations the relative number of stars, $N_{\rm CWS}/N$, which belong to the CWS. As can be seen in the middle-left panel of Fig. 4.1, this number reaches, within our model, the value of $\approx$0.5 at $t = 6$ Myr. Furthermore, Paumard et al. (2006) show that the normal vectors of the observed CWS and the CND are nearly perpendicular with $\boldsymbol{n}_{\rm CWS} \cdot \boldsymbol{n}_{\rm CND} \approx 0.01$. Hence, we investigate the evolution of the inclination $i_{\rm CWS}$ of the CWS with respect to the CND in our computations. We find that $i_{\rm CWS} \approx 90°$ at $t = 6$ Myr (see the middle-right panel of Fig. 4.1), which is in a remarkable agreement with the observational data. Finally, we investigate the eccentricity distribution n_e within the CWS and in the whole young stellar system after 6 Myr of orbital evolution. The corresponding histograms in the bottom-left panel of Fig. 4.1 show that in both cases a substantial fraction of the orbits have, in accord with the observations, moderate eccentricities. The mean eccentricity of the stars within the CWS (see the dotted line in the bottom-right panel of Fig. 4.1) is then $\approx$0.25, which is somewhat lower than the value 0.36 ± 0.06 recently reported by Bartko et al. (2009). However, at the current level of accuracy, the observations do not provide sufficient information for a reliable determination of the orbital eccentricity for a significant number of the WR/OB stars. Hence, the eccentricity criterion should be considered only as supplemental.

4.2.1 Application of the Semi-Analytic Model

In Chap. 3, we have developed a simple semi-analytic model for the orbital evolution of N mutually interacting stars on initially circular orbits that are, in addition to the dominating Keplerian potential of the central SMBH, exposed to the gravitational potential of an extended spherical cluster and a distant axisymmetric ring. We have seen that, depending on the strength of the interaction of the stars, the orbits may evolve in two qualitatively different modes. Either the orbits interact strongly and, under such circumstances, they are dynamically coupled and precess synchronously in the axisymmetric potential, or, given their mutual interaction is weaker, the orbits precess independently, periodically interchanging their angular momentum, which

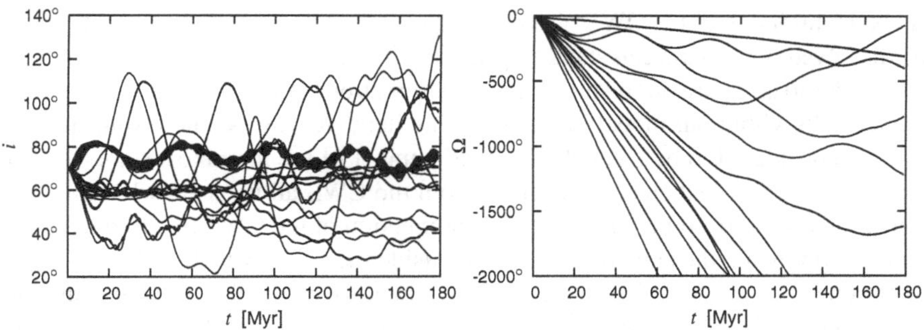

Fig. 4.3 Evolution of the initially thin stellar disc of 100 stars in the compound potential of the central SMBH, spherical cluster and axisymmetric ring. The values of orbital semi-major axes a_k in the disc range from 0.02 to 0.2 R_r and their distribution obeys $\mathrm{d}N \propto a^{-1}\mathrm{d}a$. The stellar masses are all equal with $m = 5 \times 10^{-6}\, M_\bullet$ while the mass of the ring is set to $M_r = 0.3\, M_\bullet$. Initial inclination i_0 of all the orbits with respect to the ring equals 70°

leads to oscillations of their inclinations. In the following, we shall show that this effect is responsible for the above described results of our N-body calculations.

For this purpose, we analyze the evolution of a system which bares the main qualitative features of the canonical model (see Table 4.1) by means of the previously derived Eq. (3.30). In particular, we consider an initially thin stellar disc with the distribution of semi-major axes of the orbits $\mathrm{d}N \propto a^{-1}\mathrm{d}a$. As we show in Fig. 4.3, the oscillations of the orbital inclinations no longer have the simple patterns which we observed for the models discussed in Chap. 3. On the other hand, we can still identify a well defined group of orbits which coherently change their orientation with respect to the CND towards higher inclination. These orbits thus form a rather thin disc during the whole monitored period of time. It turns out that they represent the innermost parts of the initial disc where the separations of the neighbouring orbits are small enough for their mutual interaction to couple them together. In the outer parts, however, the disc is not so dense and, therefore, the influence of the CND dominates the mutual interaction of the stars, leading to decomposition of the initially coherent structure.

Furthermore, by means of the semi-analytic formulae derived in Chap. 3, let us evaluate the order of magnitude characteristic time-scale for the stellar disc evolution. In order to determine the rough time estimate, we use formula (3.25). As this formula has been derived for a system of two stars, we replace the stellar disc from the canonical model with two characteristic particles at certain radii a', a in the sense of Sect. 3.2. For this purpose, let us divide the stars in the disc into two groups according to their initial distance from the centre and define a' and a as the radii of the orbits of the median stars in the inner and outer group, i.e. $a' = 0.06$ pc and $a = 0.23$ pc. Inserting these values into formula (3.25), we obtain $T_{\mathrm{char}} \approx 37$ Myr for the canonical model. This value is in order of magnitude agreement with the estimated age of the early-type stars, ≈6 Myr (Paumard et al. 2006), since the core of the disc reaches its maximal inclination with respect to the CND already after a fraction of period $T_{\rm char}$ as can be seen in Figs. 3.3 and 4.3. Hence, we conclude that the simple semi-analytic model developed in Chap. 3 indeed reveals the main cause of the behaviour observed in our numerical N-body calculations and represents a plausible physical explanation of the stellar configuration observed in the centre of our Galaxy.

Finally, let us also mention that, in addition to the core of the disc, less significant groups of orbits with coherent secular evolution may exist even in the outer parts of the disc (see Fig. 4.3) if their separations are small enough. Our semi-analytic approach thus admits possible existence of secondary disc-like structures in the observed young stellar system which has indeed been discussed by several authors (Genzel et al. 2003; Paumard et al. 2006; Bartko et al. 2009).

4.3 Discussion of the Problem

As we have shown in the previous section, all the WR/OB stars may have been formed within a single gaseous disc and, subsequently, brought to their present location by the effects of the dynamics in the combined potential of the central SMBH and the axisymmetric CND. In the following we establish a set of parameters for which the evolution of the young stellar system leads to a configuration compatible with the current observational data. For this purpose, we follow the evolution of the system for various values of the model parameters. Within the results, we then concentrate on $N_{\rm CWS}/N$, $i_{\rm CWS}$, $\langle e_{\rm CWS}\rangle$ and $\langle e\rangle$ and confront their values at $t = 6$ Myr with the observations.

4.3.1 Model Parameters Compatible with Observations

To begin with, we follow the evolution of the young stellar system for various values of $M_{\rm CND}$ and $M_{\rm c}$ with the other parameters fixed to their canonical values (see Table 4.1). The results indicate that the strongest constraints on the possible values of $M_{\rm CND}$ and $M_{\rm c}$ come from $N_{\rm CWS}/N$ and $i_{\rm CWS}$. Their values for $M_{\rm c} \in [0.01\, M_{\bullet}, 3\, M_{\bullet}]$ are depicted by the solid lines in the top panels of Fig. 4.4, whereas $M_{\rm CND}$ remains constant along each of the lines and is set to either $0.1\, M_{\bullet}$ or $0.3\, M_{\bullet}$ or $0.6\, M_{\bullet}$. According to these results, the evolution of the young stellar disc leads to values of $N_{\rm CWS}/N$ which accommodate the observational constraints, if $0.1\, M_{\bullet} \lesssim M_{\rm CND} \lesssim 0.3\, M_{\bullet}$ and $0.01\, M_{\bullet} \lesssim M_{\rm c} \lesssim 2M_{\bullet}$. However, the upper limit for the mass of the cluster of late-type stars has to be reduced to $M_{\rm c} \lesssim M_{\bullet}$ since larger values do not lead to the observed $i_{\rm CWS} \approx 90°$. Both observational criteria are, therefore, fulfilled if $0.1\, M_{\bullet} \lesssim M_{\rm CND} \lesssim 0.3\, M_{\bullet}$ and $0.01\, M_{\bullet} \lesssim M_{\rm c} \lesssim M_{\bullet}$. Moreover, the results of our computations with $M_{\rm c} = 0$ show that even in this case, the evolution of the young stellar system leads to a configuration which matches the observational data (due to logarithmic scale in Fig. 4.4, the corresponding values are not displayed). Hence, we

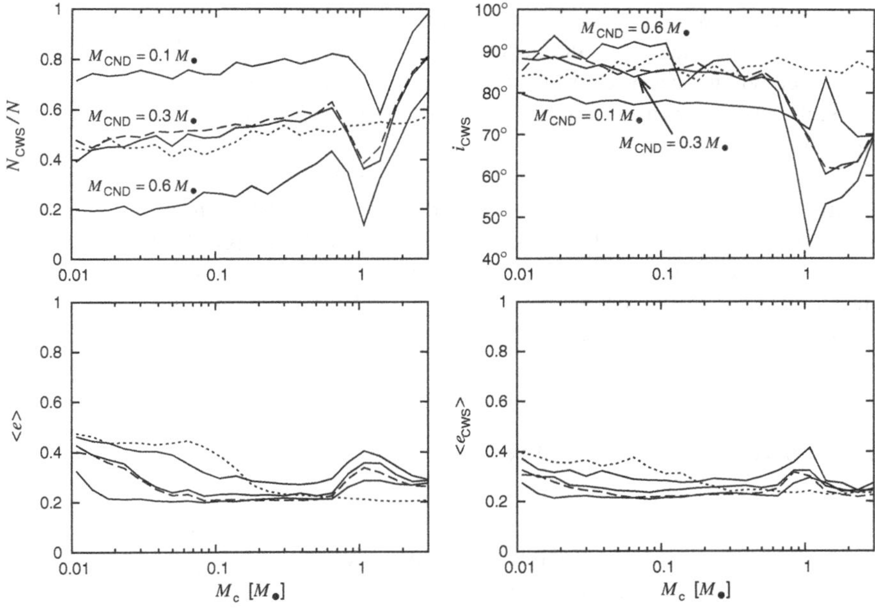

Fig. 4.4 Number of stars within the CWS (*top-left*), inclination of the CWS with respect to the CND (*top-right*) and mean eccentricity of all the stars in the young stellar system (*bottom-left*) and within the CWS (*bottom-right*) at $t = 6$ Myr for various values of M_{CND} and M_{c}. All results are averaged over 12 realizations. *Solid lines* $\beta = 7/4$, $N = 200$, $m \in [4\,M_\odot, 120\,M_\odot]$ (only properties of stars with $m \geq 12\,M_\odot$ displayed), $\alpha = 1$. *Dashed line* $M_{\mathrm{CND}} = 0.3\,M_\bullet$, $\beta = 1/2$, $N = 200$, $m \in [4\,M_\odot, 120\,M_\odot]$ (only properties of stars with $m \geq 12\,M_\odot$ displayed), $\alpha = 1$. *Dotted line* $M_{\mathrm{CND}} = 0.3\,M_\bullet$, $\beta = 7/4$, $N = 136$, single mass, $m = 50\,M_\odot$. Common parameters for all models are: $M_\bullet = 4 \times 10^6\,M_\odot$, $\Delta_0 = 2.5°$, $i^0_{\mathrm{CWS}} = 70°$

find the final intervals $0.1\,M_\bullet \lesssim M_{\mathrm{CND}} \lesssim 0.3\,M_\bullet$ and $0 \leq M_{\mathrm{c}} \lesssim M_\bullet$. If we substitute $M_\bullet = 4 \times 10^6\,M_\odot$, the intervals transform to $4 \times 10^5\,M_\odot \lesssim M_{\mathrm{CND}} \lesssim 1.2 \times 10^6\,M_\odot$ and $0 \leq M_{\mathrm{c}} \lesssim 4 \times 10^6\,M_\odot$.

The intervals for allowed M_{CND} and M_{c} are not affected if we evaluate the eccentricity criterion. As can be seen in the bottom panels of Fig. 4.4 (solid lines), all the considered values of M_{CND} and M_{c} lead to similar values of both $\langle e_{\mathrm{CWS}} \rangle$ and $\langle e \rangle$, which satisfy the observational constraints. Nevertheless, our results show that the orbital eccentricities in the young stellar system reach slightly higher values for lower M_{c}. We attribute this effect to Kozai oscillations, which are less suppressed by the cluster of late-type stars. Furthermore, our results suggest that also for $M_{\mathrm{c}} \approx M_\bullet$, all the orbits in the young stellar system gain somewhat larger eccentricities, regardless the mass of the CND. Around the same value, $M_{\mathrm{c}} \approx M_\bullet$, i_{CWS} appears to be more sensitive upon the variations of M_{c} and N_{CWS}/N reaches its minima (see the solid lines in the top panels of Fig. 4.4). Hence, it seems that all of these effects are somehow connected with a stronger influence of the CND on the dynamical evolution of the young stellar disc. However, at this point, we can not provide any explanation of this effect.

In order to investigate whether the suggested intervals for $M_{\rm CND}$ and $M_{\rm c}$ depend upon the density profile of the cluster of late-type stars, we model the evolution of the young stellar system also for $\beta = 1/2$ and $M_{\rm c} \in [0.01\,M_\bullet, 3\,M_\bullet]$. The other parameters remain at their canonical values (see Table 4.1). The dotted line in Fig. 4.4 proves that $N_{\rm CWS}/N$, as well as $i_{\rm CWS}$ and both $\langle e_{\rm CWS}\rangle$ and $\langle e\rangle$, reach the same values as in the case with $\beta = 7/4$, except for the neighbourhood of the point $M_{\rm c} \approx M_\bullet$. The absence of the 'resonant' effects observed in this case can be interpreted as a consequence of the different mass of the cluster of late-type stars enclosed within the young stellar disc, due to the different value of β. However, since the value $M_{\rm c} \approx M_\bullet$ represents only the approximate upper boundary of the suggested interval for $M_{\rm c}$, the latter effect is, for the purpose of this study, rather insignificant.

Similarly, in order to test the dependence of the intervals upon the mass function of the young stellar disc itself, we perform a set of computations with the disc treated as a group of 136 single mass stars each with mass $m = 50\,M_\odot$. The other parameters are set to their canonical values. In this case, the total mass of the disc is the same as in all the models, which we have presented so far ($\approx 6.8\times10^3\,M_\odot$). As demonstrated by the dashed line in Fig. 4.4, none of the results depend upon the mass function of the disc if its total mass is preserved.

Our calculations further show that the evolution of the young stellar disc is not affected significantly if its total mass is changed within the range $\approx 10^3$–$10^4\,M_\odot$. On this account, all the results presented in the previous section would remain entirely unaffected even if we did not include the undetected stars with mass $m \in [4\,M_\odot, 12\,M_\odot]$ as their overall mass represents only a small fraction of the total mass of the whole disc.

On the other hand, decreasing the total mass of the disc by reducing the mass of the individual stars m inhibits the effect of mutual dynamical coupling of the stellar orbits as their interaction becomes weaker. Consequently, we do not observe the evolution of $i_{\rm CWS}$ if m becomes negligibly small, i. e. if the stars in the disc can be considered as test particles. This effect is demonstrated by the right panel of Fig. 4.2, where we set $m = 0.004\,M_\odot$. We see that except for the few outermost stars, which are still slightly affected by Kozai oscillations caused by the CND, the inclination of the orbits remains constant throughout the disc. Hence, our results are in accord with the findings of Šubr et al. (2009).

We have determined the intervals for $M_{\rm CND}$ and $M_{\rm c}$ under assumption $i^0_{\rm CWS} = 70°$. Our results indicate that if $i^0_{\rm CWS} \gtrsim 60°$ and both the $M_{\rm CND}$ and $M_{\rm c}$ fall into the determined intervals, the evolution of the young stellar system leads within 6 Myr to a configuration that agrees with the current observations (see Haas and Šubr in Proceedings of RAGtime 11/12, in press). With lower values of $i^0_{\rm CWS}$ considered, the CWS is entirely destroyed by the differential precession before it can reach the orientation perpendicular to the CND. On the other hand, considering $i^0_{\rm CWS}$ closer to 90° may increase the allowed intervals for $M_{\rm CND}$ and $M_{\rm c}$. However, in order to fully understand the impact of different values of $i^0_{\rm CWS}$ on the suggested intervals, a more detailed study would be required. We will focus on this issue in our future work.

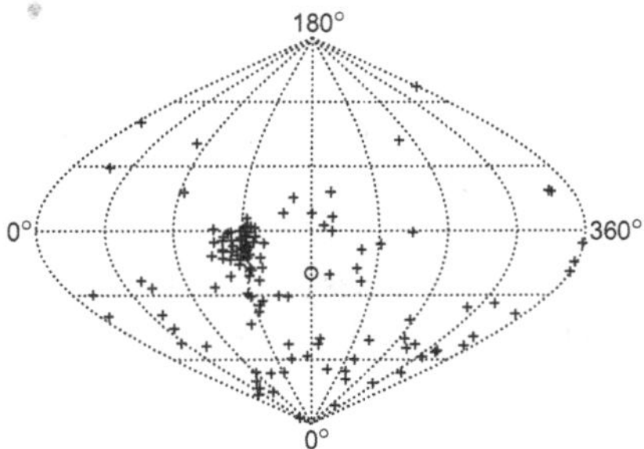

Fig. 4.5 Angular momenta of individual stars in the young stellar disc at $t = 6$ Myr (initial state denoted by an *empty circle*) within one of the realizations of the model with the CND treated as a group of $N_{\rm CND} = 20$ equal-mass particles. The other parameters are set to their canonical values (see Table 4.1), except for $M_{\rm CND} = 0.1\,M_\bullet$ and $M_{\rm c} = 0.1\,M_\bullet$

In the presented analysis, we consider the young stellar disc to be initially thin. Our calculations show that the course of its evolution does not depend upon the value of its initial half-opening angle if $\Delta_0 \lesssim 5°$.

4.3.2 Structure of the CND

So far, we have modelled the CND as a single massive particle on a circular orbit around the SMBH. This approximation has been used instead of analytical descriptions, e.g. infinitesimally thin ring, for numerical reasons. Namely, due to its simplicity, single-particle approach minimizes the necessary computational time, and the corresponding perturbing particle can be implemented into the original NBODY6 code in a trivial way.

The single-particle approximation follows from the standard averaging technique described in Sect. 1.2. As a consequence, the single-particle approximation is equivalent to the model with the CND treated as an infinitesimally thin ring if the assumptions of the averaging technique are satisfied. For the studied young stellar system, these assumptions can be written as the following two conditions for the orbital period $P_{\rm p}$ of the massive CND particle: (i) $P_{\rm p}$ must be significantly longer than the orbital periods $P_{\rm d}{}^j$ of the early-type stars in the disc, and (ii) $P_{\rm p}$ must be significantly shorter than the characteristic period $P_{\rm c}$ of the studied phenomena. Since $P_{\rm c} \sim 10^6$ yr, $P_{\rm p} \sim 10^5$ yr, and $P_{\rm d}{}^j \sim 10^2$–10^4 yr, both conditions are fulfilled and, therefore, the use of single-particle approximation is, in our case, well justified.

The real CND is, rather than a ring-like structure, a gaseous torus, which consists of several somewhat autonomous clumps (see, e.g., Christopher et al. 2005). In order to test whether the evolution of the young stellar disc can be affected by the clumpiness of the CND, we further consider the CND to be a group of $N_{\rm CND}$ equal-mass particles. It turns out that the CND constructed in this way is unstable with

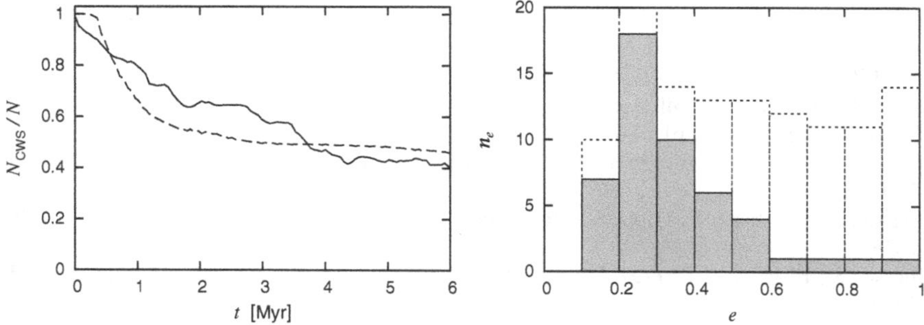

Fig. 4.6 One of the realizations of the model with the CND treated as a group of $N_{CND} = 20$ equal-mass particles. The other parameters are set to their canonical values (see Table 4.1), except for $M_{CND} = 0.1\, M_\bullet$ and $M_c = 0.1\, M_\bullet$. *Left* Number of stars within the CWS (*solid line*). For comparison, we show the results for the canonical model (*dashed line*). *Right* Eccentricity distribution at $t = 6$ Myr for all the stars in the young stellar system (*empty boxes*) and within the CWS (*grey boxes*)

respect to its own gravity. Consequently, some of the particles successively migrate towards the SMBH. These particles can eventually be identified with several gaseous streams, which are indeed observed within the radius of the CND (for one of the most recent studies, see Zhao et al. 2010). The results further show that all the effects of the dynamics in the combined potential of the central SMBH and the axisymmetric CND remain present (see Figs. 4.5 and 4.6 for the case with $N_{CND} = 20$). Moreover, the infalling CND particles pose a stronger perturbation for the young stellar disc. As a result, the orbits in the disc gain higher eccentricities compared to models with the CND treated as a single massive particle on a stable orbit (cf. the right panel in Fig. 4.6 and the bottom-left panel in Fig. 4.1). Similarly, the gradual deformation of the young stellar disc, as well as its eventual destruction, are also accelerated.

Hence, it appears that the models with the CND treated as a single massive particle somewhat underestimate its influence upon the dynamical evolution of the young stellar disc. On the other hand, the perturbative influence of the infalling parts of the gaseous CND would probably not be as strong as the impact of infalling point-like particles in the latter model. A more precise approach to the gas dynamics would thus be required in order to obtain a more accurate description of the CND.

References

Allen D. A., Hyland A. R., Hillier D. J., 1990, MNRAS, 244, 706

Bartko H. et al., 2009, ApJ, 697, 1741

Bartko H. et al., 2010, ApJ, 708, 834

Christopher M. H., Scoville N. Z., Stolovy S. R., Yun M. S., 2005, ApJ, 622, 346

Collin S., Zahn J.-P., 1999, A&A, 344, 433

Cuadra J., Armitage P. J., Alexander R. D., 2008, MNRAS, 388, L64

Do T., Ghez A. M., Morris M. R., Lu J. R., Matthews K., Yelda S., Larkin J., 2009, ApJ, 703, 1323
Eisenhauer F. et al., 2005, ApJ, 628, 246
Genzel R. et al., 2003, ApJ, 594, 812
Genzel R., Eisenhauer F., Gillessen S., 2010, RvMP, 82, 3121
Ghez A. M. et al., 2003, ApJ, 586, L127
Ghez A. M., Salim S., Hornstein S. D., Tanner A., Lu J. R., Morris M., Becklin E. E., Duchêne G., 2005, ApJ, 620, 744
Gillessen S., Eisenhauer F., Trippe S., Alexander T., Genzel R., Martins F., Ott T., 2009a, ApJ, 692, 1075
Gillessen S., Eisenhauer F., Fritz T. K., Bartko H., Dodds-Eden K., Pfuhl O., Ott T., Genzel R., 2009b, ApJ, 707, L114
Haas J., Šubr L., Kroupa P., 2011, MNRAS, 412, 1905
Haas J., Šubr L., in Proceedings of RAGtime 11/12: Workshops on Black Holes and Neutron Stars. Silesian University in Opava, Czech Republic, in press
Hobbs A., Nayakshin S., 2009, MNRAS, 394, 191
Kocsis B., Tremaine S., 2011, MNRAS, 412, 187
Levin Y., Beloborodov A. M., 2003, ApJ, 590, L33
Löckmann U., Baumgardt H., Kroupa P., 2008, ApJ, 683, L151
Paumard T. et al., 2006, ApJ, 643, 1011
Schödel R. et al., 2007, A&A, 469, 125
Šubr L., Schovancová J., Kroupa P., 2009, A&A, 496, 695
Šubr L., 2011, in Morris M. R., Wang Q. D., Yuan F., eds, ASP Conf. Ser. Vol. 439, The Galactic Center: a Window to the Nuclear Environment of Disk Galaxies. Astron. Soc. Pac., San Francisco, p. 258
Šubr L., Haas J., 2012, JPhCS, 372, 012018
Yelda S., Ghez A. M., Lu J. R., Do T., Clarkson W., Matthews K., 2011, in Morris M. R., Wang Q. D., Yuan F., eds, ASP Conf. Ser. Vol. 439, The Galactic Center: a Window to the Nuclear Environment of Disk Galaxies. Astron. Soc. Pac., San Francisco, p. 167
Zhao J.-H., Blundell R., Moran J. M., Downes D., Schuster K. F., Marrone D. P., 2010, ApJ, 723, 1097

Chapter 5
Conclusions

We have investigated the orbital evolution of an initially thin self-gravitating stellar disc in the dominating potential of the central SMBH. The potential of the SMBH has been considered Keplerian and perturbed by the potential of an extended spherically symmetric star cluster surrounding the disc. In particular, we have focused on the evolution of the root-mean-square values of eccentricity and inclination of the orbits in the disc in order to test whether it differs from the theoretical dependence e_{rms}, $i_{\mathrm{rms}} \propto t^{1/4}$ derived for isolated stellar discs. By means of numerical N-body modelling, we have shown that, given the cluster is emulated by an analytic power-law radial density profile, the evolution of the followed quantities is not affected, regardless the characteristic mass of the cluster (as long as the potential of the cluster can be considered a perturbation). This conclusion, however, changes dramatically if the cluster is modelled by a large number of gravitating stars. Our N-body calculations have revealed that, in such a case, the eccentricity vectors of the individual orbits in the cluster tend to synchronize their evolution, reaching similar orientations parallel to the disc plane. This process thus leads to formation of a somewhat flattened substructure in the cluster which is roughly perpendicular to the plane of the disc. Although the resulting deviation from the initial spherical symmetry of the cluster is rather small, our results indicate that it is strong enough to cause the Kozai-Lidov oscillations of individual stellar orbits. As the number of oscillating orbits gradually increases, the evolution of the root-mean-square values of the orbital eccentricity and inclination in the disc is accelerated. Let us point out that the large number of highly eccentric stellar orbits is important for many astrophysical processes, such as the production of hypervelocity stars and the S-stars in the Galactic Centre (e.g. Bromley et al. 2012), the tidal disruption of stars and their subsequent feeding to the SMBH (e.g. Karas and Šubr 2007) or the generation of the gravitational waves (e.g. Berry and Gair 2013).

Furthermore, it turns out that the potential of the flattened structure in the cluster induces formation of similar structures in the disc. In other words, the disc evolves a non-uniform distribution of the orbital arguments of pericentre. This feature represents an outcome of our hypothesis that can be directly confronted with the

J. Haas, *Symmetries and Dynamics of Star Clusters*,
Springer Theses, DOI: 10.1007/978-3-319-03650-2_5,
© Springer International Publishing Switzerland 2014

observational data, when these are available. In this context, the system of the early-type stars observed in the Galactic Centre is particularly promising as it is expected that, in the nearest future, the observational data will be able to provide the orbital elements for a statistically significant number of these stars.

In the second part of the thesis, we have considered the evolution of the stellar disc to be further perturbed, in addition to the spherical star cluster, by a distant axisymmetric source. In this case, we have restrained ourselves to modelling the cluster by the analytic radial density profile, which has enabled us to develop a simple semi-analytic model for the secular evolution of the orbits in the disc. Given the spherical potential is strong enough, we have shown that the evolution of initially circular orbits reduces to the evolution of inclinations and nodal longitudes. The spherical potential itself can then be factorized out from the outcoming momentum equations. Since it is not possible, in a general case, to solve the derived equations analytically, we have set up an integrator for their efficient numerical solution. The acquired results have been, in order to confirm their validity, compared with the corresponding full-fledged numerical integrations in the space of classical positions and momenta, showing a remarkable agreement. In the case of the simplest possible system of two stars on initially coplanar orbits interacting in the considered perturbed potential, we have identified two qualitatively different modes of their secular evolution. If the interaction of the stars is weak (yet still non-zero), the secular evolution of their orbits is dominated by an independent nodal precession. Difference of the individual precession rates then determines the period of oscillations of the orbital inclinations. On the other hand, when the gravitational interaction of the stars is sufficiently strong (depending on their mass and the radii of their orbits), the secular evolution of the orbits is dynamically coupled and, consequently, they precess coherently around the symmetry axis of the gravitational potential. Induced oscillations of inclinations have generally smaller amplitudes in comparison to the case of the weak interaction regime. We have further confirmed, by means of numerical integration of the derived momentum equations, that the coupling of strongly interacting orbits is a generic process that may occur even in more complex N-body systems. In particular, a subset of stars with strong mutual interaction evolves coherently and, as a result, its dynamical impact upon the rest of the N-body system is similar to the effect of a single particle of suitable mass and orbital radius.

As an example, we have investigated the orbital evolution of a disc-like structure that roughly models the young stellar system, which is observed in the Galactic Centre. It has turned out that coupling of the strongly interacting stars from the inner parts of the disc leads to their coherent orbital evolution, which allows us to observe a disc-like structure even after several million years of dynamical evolution in the tidal field of the CND. Orientation of this surviving disc then inevitably changes towards higher inclinations with respect to the CND, which is in accord with the observations. On the other hand, the stellar orbits from the outer parts of the disc evolve individually, being gradually stripped out from the parent thin disc structure. We emphasize here that the high mutual inclination of the young stellar disc and the CND, which is naturally reproduced in our model, represents an observational constraint that has not been considered in any of the previous analyses. The conclusions of our semi-analytic model have been confirmed by means of direct numerical N-body modelling of the young stellar system, considering the gravitational influence of both the analytic spherical cluster and the CND. Scanning the parameter space of our numerical model has revealed that the parameters for which the evolution of the system leads to the observed configuration are in a good agreement with the observational constraints. Hence, it appears possible for the puzzle of the origin of the early-type stars in the Galactic Centre to be solved by the hypothesis of their formation via fragmentation of a single gaseous disc, as suggested in Šubr et al. (2009), Šubr (2011) and Haas et al. (2011a, b). Our results might be more or less affected if the cluster were modelled in the full N-body way as in the first part of the thesis. Hence, we shall focus on this issue in our future work, which should result in a unified model of the innermost parsec of the Galactic Centre.

References

Berry C. P. L., Gair J. R., 2013, MNRAS, 429, 589
Bromley B. C., Kenyon S. J., Geller M. J., Brown W. R., 2012, ApJ, 749, L42
Haas J., Šubr L., Kroupa P., 2011a, MNRAS, 412, 1905
Haas J., Šubr L., Vokrouhlický D., 2011b, MNRAS, 416, 1023
Karas V., Šubr L., 2007, A&A, 470, 11
Šubr L., Schovancová J., Kroupa P., 2009, A&A, 496, 695
Šubr L., 2011, in Morris M. R., Wang Q. D., Yuan F., eds, ASP Conf. Ser. Vol. 439, The Galactic Center: a Window to the Nuclear Environment of Disk Galaxies. Astron. Soc. Pac., San Francisco, p. 258

Zeitfracht Medien GmbH
Ferdinand-Jühlke-Straße 7
99095 Erfurt, Deutschland
produktsicherheit@kolibri360.de